Studies in Systems, Decision and Control

Volume 121

Series editor

Janusz Kacprzyk, Polish Academy of Sciences, Warsaw, Poland
e-mail: kacprzyk@ibspan.waw.pl

About this Series

The series "Studies in Systems, Decision and Control" (SSDC) covers both new developments and advances, as well as the state of the art, in the various areas of broadly perceived systems, decision making and control- quickly, up to date and with a high quality. The intent is to cover the theory, applications, and perspectives on the state of the art and future developments relevant to systems, decision making, control, complex processes and related areas, as embedded in the fields of engineering, computer science, physics, economics, social and life sciences, as well as the paradigms and methodologies behind them. The series contains monographs, textbooks, lecture notes and edited volumes in systems, decision making and control spanning the areas of Cyber-Physical Systems, Autonomous Systems, Sensor Networks, Control Systems, Energy Systems, Automotive Systems, Biological Systems, Vehicular Networking and Connected Vehicles, Aerospace Systems, Automation, Manufacturing, Smart Grids, Nonlinear Systems, Power Systems, Robotics, Social Systems, Economic Systems and other. Of particular value to both the contributors and the readership are the short publication timeframe and the world-wide distribution and exposure which enable both a wide and rapid dissemination of research output.

More information about this series at http://www.springer.com/series/13304

Mohamed Abdelaziz Mohamed
Ali Mohamed Eltamaly

Modeling and Simulation of Smart Grid Integrated with Hybrid Renewable Energy Systems

 Springer

Mohamed Abdelaziz Mohamed
Electrical Engineering Department
King Saud University
Riyadh
Saudi Arabia

and

Electrical Engineering Department
Minia University
Minia
Egypt

Ali Mohamed Eltamaly
Sustainable Energy Technologies Center
King Saud University
Riyadh
Saudi Arabia

and

Electrical Engineering Department
Mansoura University
Mansoura
Egypt

ISSN 2198-4182   ISSN 2198-4190   (electronic)
Studies in Systems, Decision and Control
ISBN 978-3-319-87874-4   ISBN 978-3-319-64795-1   (eBook)
DOI 10.1007/978-3-319-64795-1

Printed on acid-free paper

This Springer imprint is published by Springer Nature
The registered company is Springer International Publishing AG
The registered company address is: Gewerbestrasse 11, 6330 Cham, Switzerland

I dedicate this book to my parents, my wife, my sons, my brothers, my sisters, my friends, and colleagues. Without their patience, understanding, support, sacrifice, and most of all love, the completion of this work would not have been possible.

Dr. Mohamed Abdelaziz Mohamed

Preface

Renewable energy power plants have been used to feed loads in remote areas as well as in central power plants connected to electric utility. Smart grid concepts used in the design process of the hybrid renewable power systems can reduce the size of components which can be translated to reduce the cost of generated energy. This book introduces a design methodology for stand-alone hybrid renewable energy system with and without applying the smart grid concepts for comparison purpose. The proposed hybrid renewable energy system contains wind, photovoltaic, battery, and diesel engine. The system is used in the beginning to feed certain loads, and the hybrid system should cover the load required completely. A novel methodology has been introduced taking the smart grid concept into account by dividing the loads into high- and low-priority parts. The high-priority part should be supplied at any generated conditions. But, the low-priority loads can be shifted to the time when the generated energy from renewable energy sources is greater than the high-priority load requirements. Results show that the using of this smart grid concept will reduce the component size and the cost of generated energy compared to the case without dividing the loads. Smart optimization techniques like particle swarm optimization (PSO) and genetic algorithm (GA) have been used to optimally design the hybrid renewable energy system.

This book will give the reader an excellent background about the renewable energy sources, optimal sizing and locating of hybrid renewable energy sources, the best optimization methodologies for sizing and designing the components of hybrid renewable energy system, and how to use smart grid concepts in the design and sizing of this system.

This book will be very interesting for the readers who look for using hybrid renewable energy system to feed loads in isolated areas. It will also help them to know about the dispatch methodology and how to connect different components of this system. Also, it will help them to understand the modeling of different components of the hybrid renewable energy system as well as the cost analysis of this system.

Riyadh, Saudi Arabia/Minia, Egypt   Mohamed Abdelaziz Mohamed
Riyadh, Saudi Arabia/Mansoura, Egypt   Ali Mohamed Eltamaly

Statement of Originality

I hereby certify that all of the work described in this book is an original work. Any published (or unpublished) ideas and/or techniques from the work of others are fully acknowledged in accordance with the standard referencing practices.

Dr. Mohamed Abdelaziz Mohamed

Acknowledgements

Thanks first and last to Allah for the utmost help and support during this work.

Thanks to King Saud University for giving me the opportunity to conduct my work and improve my scientific career.

Sincere thanks are also due to my family who suffered a lot during the preparation of this work.

Dr. Mohamed Abdelaziz Mohamed

Contents

Abbreviations[1]

RES	Renewable energy sources
HRES	Hybrid renewable energy system
WT	Wind turbines
DG	Diesel generator
HPL	High-priority load
LPL	Low-priority load
LOLP	Loss of load probability
LEC	Levelized energy cost
PSO	Particle swarm optimization
PIPSO	Parallel implementation of PSO
FREEDM	Future renewable electric energy delivery and management
NSF	National Science Foundation
RD&D	Research development and demonstration
HOMER	Hybrid optimization model for electric renewable
HYBRID2	Hybrid power system simulation model
GA	Genetic algorithm
NWT	WT number
PVA	PV area
EMS	Energy management systems
DPS	Demand profile shaping
NPSP	New proposed simulation program
NPPBSG	New proposed program-based smart grid
IOT	Iterative optimization technique
SIPSO	Serial implementation of PSO
NPPBPSO	New proposed program-based PSO
Parpool	Parallel worker pool

---

[1]Abbreviations have been arranged as appeared in the text.

Symbols[2]

SOC	Battery state of charge
SOC_{min}	Battery minimum state of charge
h	Hub height
h_g	Anemometer height
$u(h)$	Wind speed at hub height
$u(h_g)$	Wind speeds at anemometer height
α	Roughness factor
P_W	Wind turbine output power
P_r	Rated output power of wind turbine
u_c	Cutoff wind speed
u_r	Rated wind speed
u_f	Cutoff wind speed
k	Shape parameter
c	Scale parameter
$P_{WT,av}$	Wind turbine average power
C_F	Wind turbine capacity factor
$P_{L,av}$	Average annual demand
H_t	Solar radiation on tilted surface
P_{PV}	Power output of the PV system
$\mu_c(t)$	Instantaneous generating efficiency of the PV system
β_t	Temperature coefficient
μ_{cr}	Theoretical solar cell efficiency
T_{cr}	Theoretical solar cell temperature
$T_c(t)$	Instantaneous solar cell temperature at the ambient temperature
T_a	Ambient temperature
λ	The Ross coefficient
F_s	Safety factor
V_F	Factor of variability

---

[2]Symbols have been arranged as appeared in the equations.

η_{PC}	Power conditioning system efficiency
E_B	Energy of the battery bank
η_{BC}	Charging efficiency of the battery bank
η_{BD}	Discharging efficiency of the battery bank
σ	Battery self-discharge rate
$E_{B,max}$	Maximum allowable storage capacities of the battery bank
$E_{B,min}$	Minimum allowable storage capacities of the battery bank
E_{BR}	Nominal storage capacity of the battery bank
DOD	Maximum depth of discharge of the battery bank
$D_f(t)$	Hourly fuel consumption of the diesel generator
P_{Dg}	Average power per hour of the DG
P_{Dgr}	Diesel generator rated power
α_D and β_D	Coefficients of the fuel consumption curve
TPV	Total present cost of the entire system
LAE	Annual load demand
CRF	Capital recovery factor
r	Net interest rate
T	System lifetime
IC	Initial capital cost
PV_P	PV price per kW
C_{PV}	Rated power of the PV system
WT_P	Wind turbine price per kW
B_P	Battery bank price per kWh
INV_P	Inverter price per kW
DG_P	Diesel generator price per kW
OMC	Operation and maintenance cost
RC	Replacement cost
i	Inflation rate of replacement units
CRC	Capacity of the replacement units
C_U	Cost of replacement units
N_{rep}	Number of units replacements
FC	Diesel generator fuel cost
DG_h	Total operation hours of the diesel generator during the lifetime
DGc	Diesel generator cost
P_f	Fuel price per liter
PSV	Present value of scrap
SV	Value of scrap of the system components
P_{dummy}	Dummy load power
X	Vector for sizing variables
PL	Load power
P_{BC}	Battery charging power
P_{BD}	Battery discharging power
η_{inv}	Inverter efficiency
η_D	DG efficiency
E_{dummy}	Dummy load energy

PB	Battery accumulated power
HPL	High-priority load
LPL	Low-priority load
P_{LHP}	High-priority load power
P_{LLP}	Low-priority load power
P_{LLP_sum}	Amassed value of the unmet low-priority load
P_{L_low}	Accessible power used to supply low-priority load
$LOLP\_HP$	Counter for loss of load probability of HPL
$LOLP\_HP_{index}$	Designed values of the counter of $LOLP\_HP$
$P_{LL\_sum_{index}}$	Designed values of the counter of P_{LLP_sum}
$E_{dummy_{min}}$	Minimum allowable value of the dummy energy
$E_{dummy_{max}}$	Maximum allowable value of the dummy energy
$ELPL$	Low-priority load energy
$pbest_i$	Local best particle
$gbest$	Global best particle
i	Particle index
x_i	Current position of each particle
v_i	Velocity of each particle
g	Counter of generations
ω	Inertia weight factor
c_1	Self-confidence
c_2	Swarm confidence
a_1 and a_2	Uniform randomly generated numbers

List of Figures

List of Tables

Summary

In recent years, interest in renewable energy sources for power generation is progressively gaining significance in the entire world due to fossil fuel depletion, the high cost of fossil fuel, and increasing environmental concerns. Therefore, there is a big trend to use renewable energy sources to address the power generation especially for the isolated or remote areas. Utilization of many renewable energy sources with storage and backup units to form a hybrid renewable energy system can give a more economic and reliable source of energy. But, due to the nonlinear response of system components and the random nature of the renewable energy sources and load profile, the smart grid is utilized to suit and incorporate these units in order to move the power around the system as efficiently and economically as possible.

One of the most important issues in the recent studies is to optimally design the hybrid renewable energy system components to meet all load requirements with minimum cost and maximum reliability. In view of the complexity of optimization of the hybrid renewable energy systems, it was imperative to discover an effective optimization method ready to give accurate optimization results.

This book presents an optimization model to determine the optimum size of stand-alone hybrid renewable energy systems so as to meet the load requirements with possible minimum cost and highest reliability. Iterative optimization technique and particle swarm optimization algorithms (PSO) have been used through this model for seeking the optimum size of hybrid renewable energy systems, and the minimum cost of the generated energy.

Parallel implementation of PSO is a new method that has been proposed in this book to distribute and speed up the optimization process. A comparison between the results obtained from PSO algorithm and those from the iterative optimization technique has been introduced. Also, a comparison between utilizing parallel implementation of PSO and utilizing serial implementation of PSO has been presented.

Demand side management as one of smart grid applications has been presented in this book using load shifting. Load shifting has been carried out by dividing the load into two factions, high-priority load and low-priority load. High-priority load must be supplied whatever the generation conditions, while low-priority load can be shifted to be supplied during the surplus generation time of the renewable energy sources.

This study has been applied on many remote areas in different provinces of Saudi Arabia and wind turbines from different manufacturers as a case study. Also, an accurate methodology for pairing between these areas and the wind turbines has been introduced to maximize energy production and minimize the cost of generated energy.

In addition, a proposed methodology for exploitation of the dummy energy and a detailed economic methodology to obtain the cost of the generated energy have been exhibited.

The simulation results confirmed that PSO is a promising optimization technique due to its ability to reach the global optimum solution with relative simplicity and computational proficiency contrasted with the customary optimization techniques. The simulation results confirmed also that parallel implementation of PSO can save more time during the optimization process compared to the serial implementation of PSO. Furthermore, the simulation results confirmed that utilizing load shifting as one of the smart grid applications can help to get a distributed load profile, reduce the entire system cost, and reduce the CO_2 emission compared with the design without such load shifting. Finally, many valuable results have been extracted from this study that could help researchers and decision makers.

Chapter 1
Introduction and Literature Review

1.1 Introduction

Nowadays, increasing energy demand and dependence on fossil fuel become important issues facing the whole world. Therefore, there is a big trend to use renewable energy sources (RES) to address the electricity generation. High penetration of RES and energy management impose immense challenges to the power system. With these challenges, the overall system operation requires advanced technologies in order to enhance the power system performance [1]. Smart grid is a system by which the existing power system infrastructure is being upgraded with the integration of multiple technologies such as, two-way power flow, two-way communication, distributed generation, advanced automated controls, and forecasting system [2]. Smart grid system enables interaction between the generation and the consumer which allow the optimal usage of energy based on environmental, price preferences and system technical issues. This, in turn, enables the power system to be more reliable, efficient and secure [3]. Table 1.1 shows the difference between the traditional grid and the smart grid system.

The challenges that can be addressed based on smart grid system are summarized in Fig. 1.1. Energy management and demand profile shaping are important challenges of smart grid system. Energy management is the strategy of adjusting and optimizing energy, using systems and procedures so as to reduce energy requirements per unit of output while holding constant or reducing total costs of producing the output from these systems. Demand profile shaping refer to the policy of reduction or shift the electricity use during peak demand periods or according to price signals [4, 5]. This, in turn, can reduce the capacity of required standby power sources, reduce the overall plant and capital cost requirements, and also can increase the system reliability [6].

© Springer International Publishing AG 2018
M. Abdelaziz Mohamed and A.M. Eltamaly, *Modeling and Simulation of Smart Grid Integrated with Hybrid Renewable Energy Systems*, Studies in Systems, Decision and Control 121, DOI 10.1007/978-3-319-64795-1_1

Table 1.1 Comparison between the traditional grid and the smart grid

Smart grid	Traditional grid
Sensors all over	Few sensors
Digital system	Electromechanical system
Distributed generation	Centralized generation
Self-monitoring–Self-restoration	Manual monitoring–Manual restoration
Two-way communication	One-way communication
Adaptive and islanding	Failures and blackouts
Prevalence control–Remote test	Limited control–Manual test
Many customer choices	Limited customer choices

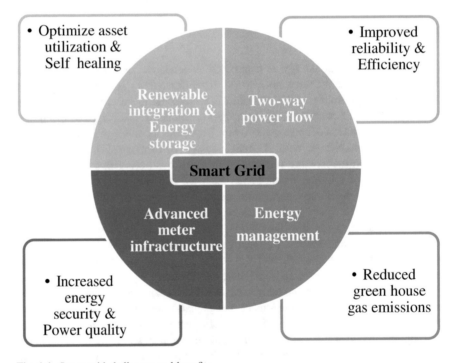

Fig. 1.1 Smart grid challenges and benefits

1.2 Problem Statement

The smart grid has recently started receiving great interest from various government organizations globally. Therefore, the smart grid and its applications are the focus of interest in this book, especially in the field of renewable energy generation. In this book, a proposed model of a hybrid renewable energy system (HRES) integrated with the smart grid is developed. The hybrid system includes wind turbines

(WT), PV arrays, diesel generator (DG), battery bank, charge controller, bidirectional converter, and load. A model for each component of the generation and load sides is introduced. The hourly data of the wind speed, solar radiation, and temperature for five sites in Saudi Arabia are used as a case study.

A novel intelligent algorithm based on smart grid applications to determine the optimal size of grid-independent hybrid PV/wind/battery/diesel energy systems so as to meet the load requirements with minimum cost and highest reliability is introduced. Load shifting-based load priority is presented in this book as one of smart grid applications by dividing the load into two factions, high priority load (HPL) and low priority load (LPL). Demand profile improvement is carried out by shifting the LPL from low generation to high generation time. The logic used in this algorithm is designed to follow the value of the loss of load probability (*LOLP*) and the dummy energy (E_{dummy}) to satisfy the aggregate load demand with a minimum value of levelized energy cost (*LEC*).

Particle swarm optimization (PSO) algorithm is employed for seeking optimum size of HRES at a minimum cost of energy of the system under study. Furthermore, parallel implementation of PSO (PIPSO) is a new proposed method introduced in this book to distribute the evaluation of the fitness function and constraints among the ready-made processors or cores and to speed up the optimization process.

1.3 Literature Review

Renewable energy integration and energy management are the major challenges for developers and practitioners of the smart grid. Therefore, these challenges were the focus of interest in this book. The following section presents a survey of the literature on integrating RES and energy management in the smart grid system.

1.3.1 Renewable Energy Integration

Renewable energy is a promising option for electricity generation, especially the wind and PV energy systems as they are clean energy sources and became mature technology. In addition, the wind energy source is considered as the world's fastest growing energy source [7] and the PV energy source is the most easily scalable type of renewable energy generation.

Recently, the studies on HRES are in terms of modeling, sizing and performance, while the studies based on integrating these systems into the smart grid are limited. Therefore, it was very important to study the integration of HRES into smart grid for the sustainable development. As an example, a survey on integrating renewable energy in smart grid system has been presented in [8]. The authors in this paper indicate the promising potential of integration of HRES into smart grid in the future and it's usefulness for developers, practitioners and policy makers of renewable energy generation. These because the use of smart grid enables to accommodate higher penetration of RES with variable cost, enhancing the system reliability, and energy efficiency.

Geviano et al. [9] surveyed and summarized the smart grid applications for renewable energy generation and its potential study in the future. The authors affirmed that the communication between the electronic devices is a key technology in order to adapt renewable energies to the future grid infrastructure.

Ayompe et al. [10] presented real-time energy models for small scale PV grid connected systems suitable for the domestic application. The models have been used to predict the real-time output power from the PV systems in Dublin, Ireland using 30 min intervals of measured data between April 2009 and March 2010.

The future renewable electric energy delivery and management system (FREEDM) was one of the smart grid applications applied on the laboratory scale and proposed by NSF at the NSF FREEDM systems center, Raleigh, NC [11]. The FREEDM system is a power distribution system that interfaces with residential and industry customers. The system operation is based on the belief that the key to avoid the energy crisis is not necessarily the renewable energy itself, but the infrastructure needed to deliver and manage large-scale distributed RES and energy storage. The objective of FREEDM is to have an efficient electric power grid integrating highly stochastic, distributed and scalable alternative generation sources and energy storage with the existing power systems.

Kohsri et al. [12] proposed an energy management and control system for smart renewable energy generation. They used LAB-View technology as a basic design for the overall system. Their proposed system is constructed as PV/wind/diesel. The system itself can forecast and make a decision for future power management.

A research development and demonstration (RD&D) micro-grid was installed in the Burnaby campus of the British Columbia Institute of Technology (BCIT) in Vancouver, British Columbia, Canada. The BCIT's smart micro-grid is a test bed where multitude of components, technologies, and applications of the smart grid are integrated to qualify the merits of different solutions, showcase their capabilities, and accelerate the commercialization of technologies and solutions for it [13]. The BCIT's smart grid development lab includes powerful servers, routers, protocol analyzers, and networking equipment and integrated with multiple base stations and smart metering installations to create an end-to-end smart grid control center. In this lab, a variety of experiments, tests, and validation efforts can be programmed and carried out.

1.3.2 Optimal Sizing of Renewable Energy Systems

The HRES used with smart grid need an accurate and optimum design and sizing to minimize the generated energy cost and maximize the system reliability. Optimum sizing of the HRES is complex and takes a long time, especially with load shifting according to priority imposed by smart grid applications. Therefore, several approaches have been developed to achieve the optimal configurations of the HRES. These approaches differ based on the optimization objectives, optimization constraints, and the used optimization techniques. As an example, the authors in [14] used a graphical construction technique for determining the optimum combination of PV array and battery in a hybrid PV/wind system. The system is simulated for various combinations of PV array, battery sizes, and *LOLP*. Then, for the desired value of *LOLP*, the PV array versus battery size is plotted and the optimal solution, which minimizes the total system cost, can be chosen. This type of graphical technique doesn't allow including more than two parameters in the optimization process (i.e. PV/wind, wind/battery, or PV/battery).

A linear programming technique has been introduced in [15] for optimum sizing of hybrid PV/wind/battery power system to feed a certain load with minimum system total cost. The total system cost consists of the initial cost, the running cost, and the maintenance cost.

Belfkira et al. [16] reported the practical interest of using the sizing methodology and showed the effect of the battery storage on the total cost of the HRES. These authors presented a deterministic algorithm that minimizes the total system cost while satisfying the load requirements of a stand-alone hybrid PV/wind/diesel energy system. This algorithm uses six months data of wind speed, solar radiation, and temperature. The main disadvantage of this algorithm is using a short-term meteorological data of wind speed, solar radiation, and temperature which reduces the system sizing accuracy. In addition, this algorithm did not apply the smart grid applications when determining the optimum size of the system under study.

A study was introduced in [17] for optimally sizing an autonomous hybrid PV/wind energy system with battery storage on techno-economic basis. The level of autonomy and the cost of the system were the targeted objectives of this study. The author in this study used a numerical analysis based on 1994 weather data from the TyB experimental site of the Cardiff University. The main advantage of this study is it considers the monthly variation in the required size of HRES to overcome the limitations of the scenarios which select the optimum size based on the worst renewable month(s). However, this study did not take the smart grid applications into account.

An iterative optimization technique was introduced in [18] for the techno-economic optimization of a hybrid PV/wind energy system with/without an uninterruptible power supply to supply certain load. The optimum size of HRES components and the lowest *LEC* were the main optimization objectives. The authors compared the performances of HRES with and without the uninterruptible power supply and reported that the type of system configuration affects *LEC* and the battery state of charge (*SOC*), especially at low windy sites. Furthermore, the authors confirmed that the hybrid system is the best option for the systems under study. However, the authors did not use the smart grid applications while determining the optimum system size.

Kaabechea et al. [19], Yang et al. [20] and Kellogg et al. [21] presented a sizing optimization model for a hybrid PV/wind/battery energy system using the iterative technique. This model takes the *LOLP* and the *LEC* as the optimization objectives in determining the optimum system configuration. The author in [21] introduced several possible combinations of PV/wind generation capacities and the one with the lowest cost was selected. The authors in [19–21] neglected the application of the smart grid applications in the proposed optimization model. Cost calculations of the diesel have not been included in [21] which in turn reduce the calculation accuracy.

An iterative optimization methodology has been introduced in [22] to optimize the size of a hybrid PV/wind/battery energy system subject to minimum investment cost of the system components. This methodology first assumes that the battery capacity is infinite to determine the maximum battery and minimum supply size, and then estimates the optimum number of WT and PV arrays that supply the load demand with a given *LOLP*. The authors considered the effect of utilizing real meteorological data when contrasted with the utilization of generic analytical meteorological models for sizing the system components. The authors affirmed that the use of meteorological models yielded more precise system sizes as compared to the use of real data. Be that as it may, the authors did not utilize the smart grid applications while deciding the optimum system size.

A research work has been carried out by Nandi and Ghosh [23] on optimization of a hybrid PV/wind/battery system and its performance for a typical community load in Bangladesh. For feasibility and optimum sizing of the system components, the hybrid optimization model for electric renewable (HOMER) was used. The major disadvantage of HOMER software is that the sizing of the system components assumes many simplifications during the optimization process. These simplifications have a significant impact on the accuracy of results deduced from the HOMER. In addition, the analysis requires more information on resources, economic constraints, and control methods which might be hard to be accomplished.

Barley et al. [24] introduced a description and analysis of hybrid wind/PV system for providing electricity to about one-third of the non-grid connected households in Inner Mongolia. They applied Hybrid power system simulation model (HYBRID2) software in conjunction with a simplified time-series model. Sizing of the major components of the system was determined based on the trade-off between the cost of the system and the percent unmet load. The result showed that using PV to the wind system in conjunction with battery storage reduces the unmet load by over 75%. HYBRID2 can simulate the hybrid systems with remarkably high precision calculations, but it does not optimize the size of the system components.

Eltamaly et al. [25] addressed the economic sizing of HRES using three sites of Saudi Arabia as a case study. They introduced a methodology to determine the best wind turbine, WT type out of 140 wind turbines from different manufacturers and the best-fit WT for a site to maximize energy production. Similarly, the best-fit WT is matched with the best-fit PV arrays in a determined penetration ratio to meet the load requirements of the sites under study. These authors designed the system with the assumption of zero *LOLP*.

In [26] the authors proposed a methodology which suggests using commercially available devices for their design. They used Genetic Algorithm (GA) to determine the optimal number and type of units for stand-alone PV/wind generation systems.

It ensures that the 20-year round total system cost is minimized, subject to the constraint that the load energy requirements are completely covered.

Katsigiannis et al. [27] used GA to optimize hybrid energy systems which consisted of PV/WT/battery/diesel generator as a secondary energy source to supply three isolated islands in Japan. The proposed method determines the optimum number of PV arrays, WT and battery banks based on minimum total investment cost.

The authors in [28] recommended an optimal design model utilizing GA for designing hybrid PV/wind/battery systems. The model is employed for calculating the system optimum configurations and ensuring that the annualized cost of the systems is minimized while satisfying the custom required *LOLP*. The decision variables included in the optimization process are the PV module number, PV module tilt angle, WT number (*NWT*), WT installation height and the number of batteries.

Mellit in [29] presented a methodology using GA combined with neural network algorithm for the optimization of a hybrid PV/wind/battery energy system. PV area (*PVA*), battery capacity, and WT capacity were the optimization parameters and the system cost was the objective function. The author in this methodology used a constant load for the system which is considered inaccurate assumption and is away from the real shape of loads.

In the last decade, numerous authors developed PSO to fulfill several HRES optimization objectives and constraints [30–40]. As an example, Boonbumroong et al. [30] utilized PSO to minimize the life-cycle cost of a stand-alone PV/wind/diesel system to feed a certain load. The optimization constraint was that the hourly energy demand must be satisfied by the amount of generated energy.

A PSO algorithm was applied in [31–37] for optimum sizing of a hybrid energy system supplying a certain load. The optimization objective was to minimize the system cost with the constraint of having specific reliability. In [36], maximizing the total net present worth was the optimization objective. The hybrid system includes PV/wind/battery in [31, 33, 35–37], PV/wind/fuel cell in [32], and PV/wind/tidal/battery energy sources in [34].

Hakimi et al. [38] used PSO to minimize the total cost of a stand-alone hybrid energy system formed by wind units, electrolytes, a reformer, an anaerobic reactor, fuel cells and some hydrogen tanks such that the demand is met. The optimization constraint was the stored energy in hydrogen tanks.

An optimization problem using PSO to solve the PV/wind capacity coordination for a time-of-use rate industrial user was introduced in [39] with the aim of maximizing the economic benefits of investing in the wind and PV generation systems. The optimization constraint was that the generation from RES must not to be greater than the maximum annual load demand.

Wang et al. [40] used multi-objectives PSO algorithm to optimize a hybrid PV/wind/battery energy system on the basis of cost, reliability, and emission criteria without considering load management.

1.3.3   Energy Management Systems

With the framework of smart grid and energy management systems (EMS), many management challenges become possible. Demand profile shaping (DPS) is the important challenge of the EMS. DPS can be accomplished by shifting, scheduling, or reducing demand in order to obtain a smoothed demand profile, or reduce peak demand. As an example, Cui et al. [41] proposed a dynamic pricing framework providing incentives to users to create full load profile appropriate for them and utilities, and almost approaching to an ideal flat profile.

Conejo et al. [42] described an optimization model to adjust the hourly load level in response to hourly electricity prices. The objective of the model is to maximize the utility of the consumer subject to a minimum daily energy consumption level, maximum and minimum hourly load levels, and ramping limits on such load levels. The authors also modeled the price uncertainty using robust optimization techniques.

Caron and Kesidis [43] proposed a dynamic pricing scheme incentivizing consumers to achieve an aggregate load profile suitable for utilities and studied how close they can get to an ideal flat profile depending on how much information they share. In addition, they provided distributed stochastic strategies that successfully exploit this information to improve the overall load profile when users have only access to the instantaneous total load on the grid.

Kishore and Snyder [44] presented a simple optimization model for determining the timing of appliance operation to take advantage of lower electricity rates during off-peak periods. They proposed a distributed scheduling mechanism to reduce peak demand within a neighborhood of homes. Their mechanism provides homes a guaranteed base level of power and allows them to compete for additional power to meet their needs. Finally, they introduced more powerful energy management controller's optimization model, based on dynamic programming, which accounts for the potential for electricity capacity constraints.

Mohsenian-Rad and Leon-Garcia [45] proposed an optimal and automatic residential energy consumption scheduling framework, which attempts to achieve the desired trade-off between customers who are more willing to reduce their aggregate demand over the entire horizon, rather than shifting their load to off-peak periods, which tend to receive higher incentives, and vice versa.

Ghosh et al. [46] developed an optimization mechanism incentivizing the energy customers. This mechanism depends on the trade-off between minimizing the electricity bills and minimizing the waiting time for the operation of each device.

Lee et al. [47] first developed a generalized measure of dispatchability of energy, identified two classes of dispatchable energy loads, and then they created models for these loads to match their consumption to the generation of energy sources.

O'Neill et al. [48] presented a novel energy management system for residential demand response. The algorithm named CAES and used to reduce the residential energy costs and smooth the energy usage. CAES is an online learning application that implicitly estimates the impact of future energy prices and consumer decisions on long-term costs and schedules residential device usage.

Emission control is an important management objective in the electric power industry and has a significant influence on environment protection. Therefore, many researchers have investigated how to optimize emission reduction. Bakker et al. [49] presented a three-step control strategy to optimize the overall energy efficiency and increase generation from renewable resources with the ultimate goal to reduce the CO_2 emission caused by electricity generation.

Saber and Venayagamoorthy [50] presented cost and emission reductions in a smart grid by maximum utilization of gridable vehicles (GVs) and RES. They presented possible models for GV applications, including the smart grid model; these models offer the best potential for maximum utilization of RES to reduce cost and emission from the electricity industry.

A common disadvantage of the optimization methods described above is that they focused on one problem like sizing, matching, reliability, emission control or cost minimization, and they didn't address multi-objectives or multi-constraints analysis of HRES [14–39]. Also, in some researches, the minimization of the system cost is implemented by using probability programming techniques or by linearly changing the values of corresponding decision variables, resulting in suboptimal solutions and sometimes increased the computational effort requirements [15–17]. Furthermore, some of these methods [14, 20, 28] did not take into account some system design characteristics, such as PV modules slope angle and WT installation height, which highly affect the optimization accuracy. The choice of the appropriate configuration of the proposed system depends on the type of output power of most generation systems and load types. However, most of the proposed systems described in [14, 16, 17–24, 26–28, 30, 32, 35–39] used a DC-bus alone or AC-bus alone for HRES despite the difference in generated energy sources. Furthermore, all of the algorithms developed in the above literature do not consider the management and exploitation of the dummy energy while renewable generation greater than the load demand and the maximum acceptable power in storage devices [14–50].

Concerning EMS and DPS in the smart grid system, some of the optimization approaches described in the above studies didn't consider load shifting and management-based load priority, and management-based available generation [14–50]. Also, some of these approaches relied on the consumer's endeavors or encouragement consumers to improve the load profile or decrease peak demand, which makes it hard to be accomplished [41, 43, 45, 46, 48]. Furthermore, most of these approaches depend on the presence of a real-time pricing tariff for cost or peak demand reduction which is sometimes not available [42, 44, 47]. Cost reduction or peak demand reduction can be accomplished by applying smart grid applications like load shifting based-available generation, which has been considered in this book.

This book introduces a methodology to determine the optimum design of the HRES. Load shifting based-smart grid application and PSO algorithm have been utilized together in this methodology, which has not been reported before in the literature, and considered as the main new contributions in this work.

1.4   Research Objectives

The main objective of this study is to model, optimize and simulate the HRES integrated with the smart grid. This study covers the following objectives:

- Developing a mathematical model for the HRES components.
- Developing a model for the economic calculations of the proposed system.
- Developing a control strategy for the demand based generation.
- Optimizing the size of the proposed system, based on smart grid application.
- Developing a proposed methodology for management and exploitation of the dummy energy.

1.5   Organization of Book

The book is organized in six chapters including this chapter. The following is a brief description of each chapter:

In this chapter introduces an overview of renewable energy generation. The challenges that can be addressed based on the smart grid are presented. In addition, the problem statement, the research objectives, and the book outline are presented. Finally, a survey of the literature on integrating RES and energy management in the smart grid system is presented.

Chapter 2 introduces a model for the proposed HRES. A mathematical model for each part of the HRES is presented. System reliability model and energy cost model for HRES are also introduced.

Chapter 3 introduces a new iterative optimization algorithm for sizing the components of a stand-alone hybrid PV/wind/diesel/battery energy system to meet the load demand with the lowest generated energy cost and the highest reliability.

Chapter 4 introduces a novel intelligent algorithm to determine the optimum size of stand-alone HRES using smart grid applications. Demand profile shaping is introduced using the load shifting method.

Chapter 5 introduces a PSO algorithm based on smart grid applications to determine the optimum size of stand-alone HRES so as to meet the load requirements with the minimum cost and the highest reliability. Parallel implementation of PSO is introduced to distribute and speed up the optimization process.

Chapter 6 provides general conclusions of this book.

Chapter 2
Modeling of Hybrid Renewable Energy System

2.1 Introduction

RES can be connected together in a DC-bus, or AC-bus, or in a hybrid DC/AC-buses. The choice of the appropriate configuration depends on the type of output power for most generation and loads. Therefore, it is better to use DC-bus coupling if most generation and some loads are DC [51] and to use AC-bus coupling in the case of mainly AC generation and loads [52]. If the major power sources of the HRES generate a mixture of AC and DC power, then a hybrid-coupled integration scheme is preferable (i.e. hybrid DC/AC-buses) [53], which is the case considered in this book as shown in Fig. 2.1.

2.2 Modeling of Hybrid PV/Wind/Battery/Diesel Energy System

The configuration used in Fig. 2.1 consists of wind energy and PV energy systems, DG, battery bank, charge controller, bidirectional converter, main load, and dummy load. The dispatch of this configuration is easy to be understood. The main load is supplied primarily from the WT and PV array through the bidirectional converter. The excess power from the wind energy system and/or PV energy system above the load demand is stored in the battery bank until the batteries are completely charged. If the battery storage is full; excess power (i.e., dummy power) will be used to supply certain special loads (i.e., dummy loads), such as loads for cooling and heating purposes, water pumping, and charging the batteries of emergency lights. When the load power is greater than the generated power, the deficit power will be compensated from the batteries until they reach the minimum SOC (SOC_{min}). When the battery storage is exhausted and the HRES fails to meet the load demand, DG is

© Springer International Publishing AG 2018
M. Abdelaziz Mohamed and A.M. Eltamaly, *Modeling and Simulation of Smart Grid Integrated with Hybrid Renewable Energy Systems*, Studies in Systems, Decision and Control 121, DOI 10.1007/978-3-319-64795-1_2

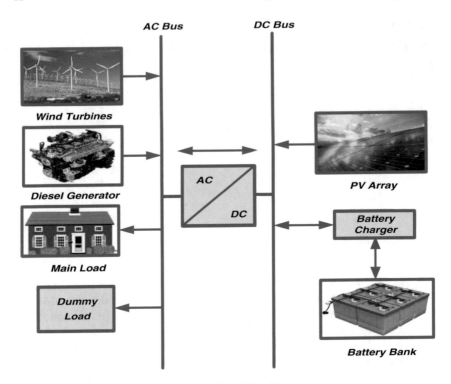

Fig. 2.1 Schematic diagram of the hybrid PV/wind/diesel/battery energy system

used. Mathematical modeling of the proposed HRES parts is detailed in the following subsections.

2.2.1 Modeling of Wind Energy System

Wind resources and the electric power output from WT at a particular location depend on wind speed at the hub height, the WT speed characteristics. Wind speed at the hub height of WT is calculated by the power law equation using the wind speed data collected at the anemometer height as [54]:

$$u(h) = u(h_g)\left(\frac{h}{h_g}\right)^{\alpha} \tag{2.1}$$

where, $u(h)$ and $u(h_g)$ are wind speeds at hub height (h) and anemometer height (h_g), respectively, and α is the roughness factor. The value of α differs from site to site and from time to time at the same site and has been taken in this book as 0.14 [55].

The output power of WT is described in terms of wind speed from the typical power curve characteristics of the WT as follows [56]:

$$P_W(u) = \begin{cases} 0, & u < u_c \text{ or } u > u_f \\ P_r \frac{u^2 - u_c^2}{u_r^2 - u_c^2}, & u_c \leq u \leq u_r \\ P_r, & u_r \leq u \leq u_f \end{cases} \tag{2.2}$$

where, P_W is the WT output power, P_r is the rated output power of WT, u_c is the cut-in wind speed, u_r is the rated wind speed, and u_f is the cut-off wind speed.

The capacity factor of the WT can be calculated as follows:

$$C_F = \frac{\exp\left[-(u_C/c)^k\right] - \exp\left[-(u_r/c)^k\right]}{(u_r/c)^k - (u_C/c)^k} - \exp\left[-(u_f/c)^k\right] \tag{2.3}$$

Weibull distribution is a statistical tool that can be used to model wind speeds. This tool can identify how often winds of different speeds will be seen at a particular location with a certain average wind speed. The Weibull parameters, shape parameter (k) and scale parameter (c) are calculated using the following statistical analysis method, respectively [57]:

$$k = a$$
$$c = \exp(-b/k) \tag{2.4}$$

where,

$$a = \left(\sum_{i=1}^{w} (x_i - \bar{x}) \sum_{i=1}^{w} (y_i - \bar{y}) \right) / \sum_{i=1}^{w} (x_i - \bar{x})^2 \tag{2.5}$$

$$b = \bar{y}_i - a\bar{x}_i = \frac{1}{w} \sum_{i=1}^{w} y_i - \frac{a}{w} \sum_{i=1}^{w} x_i \tag{2.6}$$

$$y_i = \ln(-\ln(1 - F(u_i))), \quad x_i = \ln(u_i) \tag{2.7}$$

where, $\bar{x}$ and $\bar{y}$ are the mean values of x_i and y_i, respectively.

The average power generated by each WT at a certain site can be calculated in terms of the Weibull parameters and the capacity factor from the following equations:

$$P_{WT,av} = C_F \times P_r \tag{2.8}$$

where, $P_{WT,av}$ is WT average power.

The number of WT (*NWT*) required to supply an average annual demand ($P_{L,av}$) can be calculated by the following equation:

$$NWT = \frac{P_{L,av}}{P_{WT,av}} \tag{2.9}$$

2.2.2   Modeling of PV Energy System

The solar radiation on tilted surface (H_t) can be estimated considering the solar insolation, ambient temperature, and manufacturer's data of the PV panels, slope of the PV panels and latitude and longitude of the site [58, 59]. The output power of the PV system (P_{PV}) is calculated as expressed in the following equation [60]:

$$P_{PV}(t) = H_t(t) \times PVA \times \mu_c(t) \tag{2.10}$$

where, $\mu_c(t)$ is the hourly generating efficiency of the PV system and can be obtained in terms of the cell temperature as shown in the following equation [60]:

$$\mu_c(t) = \mu_{cr}[1 - \beta_t \times (T_c(t) - T_{cr})] \tag{2.11}$$

where, β_t is the temperature coefficient, ranging from 0.004 to 0.006 per °C for silicon cells [61]. μ_{cr} and T_{cr} are the theoretical solar cell efficiency and temperature at solar radiation flux of 1000 W/m², respectively. In this book, β_t has been taken as 0.004 per °C. μ_{cr} and β_t are usually given by the PV module manufacturers. For the usual theoretical temperature $T_{cr} = 25$ °C, a literature average value for crystalline silicon modules theoretical efficiency is $\mu_{cr} = 0.12$. $T_c(t)$ is the hourly solar cell temperature at the ambient temperature (T_a), and can be obtained from the following equation [61]:

$$T_c(t) = T_a + \lambda H_t(t) \tag{2.12}$$

where, λ is the Ross coefficient, expresses the temperatures rise above ambient with increasing solar flux. Earlier reported values for λ were in the range 0.02–0.04 Cm²/W [61]. The value of λ has been used in this book as 0.03 Cm²/W.

PVA is the total solar cells area required to supply the load demand and can be calculated from the following equation:

$$PVA = \frac{1}{8760} \sum_{t=1}^{8760} \frac{P_{L,av}(t)\, F_s}{H_t\, \eta_c(t)\, V_F} \tag{2.13}$$

where, F_s is the safety factor which includes the possible allowance of insolation data inaccuracy, V_F is the factor of variability which considers the impact of yearly radiation variation, and their values are around 1.1 and 0.95, respectively.

2.2.3 Battery Storage Model

The *SOC* after certain time (t) is calculated based on the energy balance between the wind, PV energy systems and the load as given by the following equations:

$$E_B(t+1) = E_B(t)(1 - \sigma) + surplus\,power \times \eta_{BC} \quad \text{Charging mode} \quad (2.14)$$

$$E_B(t+1) = E_B(t)(1 - \sigma) - deficit\,power/\eta_{BD} \quad \text{Discharging mode} \quad (2.15)$$

where, E_B is the energy of the battery bank, η_{BC} and η_{BD} are the charging and discharging efficiency of the battery bank (in this book η_{BC} and η_{BD} have been considered as 90% and 85%, respectively) [62]. σ is the battery self-discharge rate; it is assumed as 0.2% per day for most batteries [63].

At any time, the battery bank should follow the following constraints:

$$E_{B,\min} \leq E_B(t) \leq E_{B,\max} \quad (2.16)$$

$$E_B(t+1) = E_B(t)\,(1 - \sigma) \quad (2.17)$$

where, $E_{B,max}$ and $E_{B,min}$ are the maximum and minimum allowable storage capacities of the battery bank, respectively. $E_{B,min}$ can be obtained from the following equation:

$$E_{B,\min} = DOD \quad E_{BR}, \quad (2.18)$$

where, E_{BR} is the nominal storage capacity of the battery bank, and *DOD* is the maximum depth of discharge of the battery bank.

2.2.4 Diesel Generator Model

DG is the conventional source of energy which is used as a backup to supply the power deficiency in HRES. The hourly fuel consumption of DG is assessed using the following equation [64]:

$$D_f(t) = \alpha_D P_{Dg}(t) + \beta_D P_{Dgr} \quad (2.19)$$

where, $D_f(t)$ is the hourly fuel consumption of DG in L/h, P_{Dg} is the average power per hour of the DG, kW, P_{Dgr} is the DG rated power, kW, α_D and β_D are the coefficients of the fuel consumption curve, L/kWh, these coefficients have been considered in this book as 0.246 and 0.08145, respectively [65].

2.3 System Reliability Model

In this book, the reliability of the HRES is developed based on the concept of *LOLP* which is considered as the technical implemented criteria for sizing HRES and can be defined as [66]:

$$LOLP = \frac{\sum_0^t \text{Deficit Load Time}}{8760} \, 100\% \tag{2.20}$$

2.4 Energy Cost Model

LEC is a standout amongst the most well-known and utilized indicator of economic analysis of HRES and it can be calculated using the following equation [67]:

$$LEC = \frac{TPV \times CRF}{LAE} \tag{2.21}$$

where, *TPV* is the total present cost of the entire system, *LAE* is the annual load demand, and *CRF* is the capital recovery factor. *CRF* and *TPV* are expressed as:

$$CRF = \frac{r(1+r)^T}{(1+r)^T - 1} \tag{2.22}$$

$$TPV = IC + OMC + RC + FC - PSV \tag{2.23}$$

where, *r* is the net interest rate (the interest rate for the genuine monetary condition in Saudi Arabia is 2% [68]), and *T* is the system lifetime in years (the system lifetime has been chosen in this book as 25 years).

IC is the initial capital cost of the HRES components, including the civil work, installation cost, and electrical connections and testing. In this study, the civil work and installation costs have been taken as 40% of PV generator price for the PV part and 20% of wind generator price for the wind part [69]. IC can be determined from the following equation:

$$IC = 1.4 \times PV_P \times C_{PV} + 1.2 \times WT_P \times P_r \times NWT + E_{BR} \times B_P + P_{inv} \\ \times INV_P + P_{Dgr} \times DG_p \tag{2.24}$$

where, PV_P is the PV price per kW ($/kW), C_{PV} is the rated power of the PV system (kW), WT_P is the WT price per kW ($/kW), E_{BR} is the battery capacity (kWh), B_P is the battery bank price per kWh ($/kWh), INV_P is the converter price per kW ($/kW), and DG_P is the DG price per kW ($/kW).

OMC is the present value of operation and maintenance cost of the HRES segments all over the lifetime of the system. OMC include such items as an operator's salary, inspections, insurance and all scheduled maintenance. Some researchers used a fixed percentage of the total cost of the system for maintenance, For example, in [70]; the annual maintenance cost has been set at 5% of capital cost for the WT, 1% of capital cost for the PV generator, and 0% for the batteries storage. Also, in [71] OMC has been assumed to be 1% of the initial hardware system cost. OMC has been used as a fixed cost per capacity of each component of the HRES such as in [72] the annual maintenance cost of WT has been set as $100/kW which is about 3% of the WT price and 0% of the PV system. Reference [63] used annual maintenance cost of WT as $20/kW and $10/kW for the PV system and $25/kWh of the battery capacity. Reference [73] used annual maintenance cost to be $20/kW for PV, $75/kW for WT, and $20/kWh of the capacity of the batteries. In this study, after a detailed survey of a lot of researches [70–73], the predicted value of the OMC cost is summarized as shown in Table 2.1. OMC can be determined using the following equations [74]:

$$OMC = OMC_0 \left(\frac{1+i}{r-i}\right)\left(1 - \left(\frac{1+i}{1+r}\right)^T\right) \quad r \neq i \tag{2.25}$$

$$OMC = OMC_0 \times T \quad r = i \tag{2.26}$$

where, OMC_0 is the operation and maintenance cost at the first year of the project lifetime.

Table 2.1 The economic and technical parameters of the HRES components

Item	Price ($)	Replacement cost ($)	Lifetime years	OMC (%)	Scrap value (%)	Number of replacements	Salvage times
WT, kW	3000	2400	20	3	20	1	2
Civil work For wind, kW	(20%) 600	20%	25	3	20	0	1
PV, kW	2290	2000	25	1	10	0	1
Civil work For PV, kW	(40%) 916	40%	25	1	20	0	1
Converter, kW	711	650	10	Null	10	2	3
DG, kW	850	850	10	3	20	2	3
Batteries, kWh	213	170	4	3	20	6	7

RC is the present value of replacement cost of the HRES components occurring throughout the system lifetime and can be determined as follows [69]:

$$RC = \sum_{j=1}^{N_{rep}} \left(C_{RC} \times C_U \times \left(\frac{1+i}{1+r} \right)^{T*j/(N_{rep}+1)} \right)$$ (2.27)

where, i is the inflation rate of replacement units (the inflation rate in Saudi Arabia is 2.3% [75]), C_{RC} is the capacity of the replacement units (kW for the WT, PV array, DG and inverter, and kWh for the battery bank), C_U is the cost of replacement units ($/kW for the WT, PV array, DG and inverter, and $/kWh for the battery bank), and N_{rep} is the number of units replacements over T.

FC is the DG fuel cost and can be calculated from the following mathematical statement:

$$FC = D_f(t)\, DG_h\, P_f$$ (2.28)

where, DG_h is the total operation hours of the DG during T and P_f is the fuel price per liter ($/L), (fuel price has been considered in this book by 0.8 $/L).

PSV is the present value of scrap (i.e. the system's net worth in the final year of its lifetime period). This value has been taken as 10% of WT and civil work while its value for other components has been ignored in [76]. In [71] PSV has been assumed to be 20% of the power conditioning equipment and the battery bank, and 10% for the solar array. In this study, the analysis assumes the scrap or salvage value (SV) of each component as 20% for WTs, batteries, power conditioners, and civil work and 10% for the solar array as shown in Table 2.1. PSV can be expressed by the following [77]:

$$PSV = \sum_{j=1}^{N_{rep}+1} SV \left(\frac{1+i}{1+r} \right)^{T*j/(N_{rep}+1)}$$ (2.29)

Table 2.2 The technical characteristics of the WT used in this study

WT No.	Manufacturer	Pr (kW)	D (m)	u_c (m/s)	u_r (m/s)	u_f (m/s)	H (m)
WT 1	Enercon-1	330	34	3	13	34	50
WT 2	ACSA-1	225	27	3.5	13.5	25	50
WT 3	Fuhrlander-3	250	50	2.5	15	25	42
WT 4	Ecotecnia-2	600	44	4	14.5	25	45
WT 5	ITP-1	250	30	3	12	25	50
WT 6	NEPC-3	400	31	4	15	25	36
WT 7	Southern Wind Farms	225	29.8	4	15	25	45
WT 8	Enercon-2	330	33.4	3	13	34	37
WT 9	NEPC-2	250	27.6	4	17	25	45
WT 10	India Wind Power	250	29.7	3	15	25	50

The economic and technical parameters of HRES components are shown in Table 2.2 [78, 79]. Table 2.2 summarizes the initial costs, operation and maintenance costs, replacement costs, scrap values, and lifetime of each component of HRES. The cost of each component is based on the prices installed in the recent market.

2.5   Resources and Load Data

The hourly data of the wind speed, solar radiation, and temperature for five sites in Saudi Arabia are used as a case study in this book. These data have been obtained from the King Abdulaziz City for Science and Technology (KACST). These sites are Yanbu, Dhahran, Dhalm, Riyadh, and Qaisumah [80]. These sites represent the climatic conditions variety in Saudi Arabia with different solar radiation, wind speed potentials, and temperature. Yanbu is a major Red Sea port in Al Madinah province at the west of Saudi Arabia. It is around 300 km northeast of Jeddah at (24°05′ N 38°00′ E). Dhahran is situated in the eastern part of Saudi Arabia close to the Arabian gulf coast and just a few blocks south of Dammam at (26°16′ N 50°09′ E). Dhalm is situated in the east of Taif and just about 230 km rounded at (22°43′0″ N 42°10′0″ E). Riyadh is the capital and largest city of Saudi Arabia. It is situated in the center of the Arabian Peninsula on a large plateau at (24°38′ N 46°43′ E). Qaisumah is a

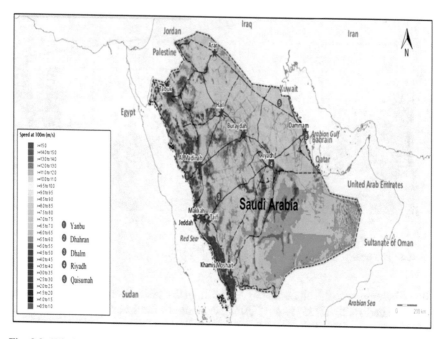

Fig. 2.2 Wind speed map in Saudi Arabia at the height of 100 m

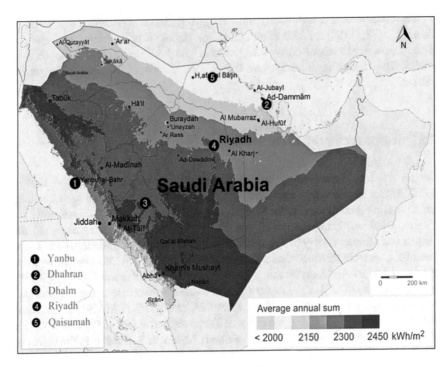

Fig. 2.3 Global horizontal radiation map in Saudi Arabia

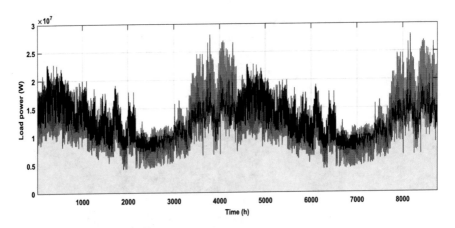

Fig. 2.4 The hourly load demand

village belonging to the city of Hafar Al-Batin, in the eastern province, Saudi Arabia
and is located at (28°18′35″ N 46°7′39″ E). Modified wind speed and horizontal solar
radiation maps of these sites are shown in Figs. 2.2 and 2.3, respectively [81].

Ten WT from diverse producers are used in this book as the wind generator, the technical characteristics of the WT under study are shown in Table 2.2 [82].

A load demand of Addfa city in Al Jouf province is used for the system under study and has the hourly demand as shown in Fig. 2.4. This load is assumed to be the same in each site.

Chapter 3
Sizing and Techno-Economic Analysis of Stand-Alone Hybrid Photovoltaic/Wind/Diesel/Battery Energy Systems

3.1 Introduction

Most of the remote areas in Saudi Arabia depend on conventional electric energy sources such as the DG. However, these sources depend on the availability of the expensive fossil fuel. Also, these engines usually operate at low efficiency owing to their different loads. Therefore, using RES that doesn't need fuel and doesn't affect the environment can be considered an excellent solution for this dilemma. Using various sources of renewable energy can increase system reliability and reduce the cost of generating energy considerably.

This chapter proposes a new iterative optimization algorithm for sizing the components of the system described in Fig. 2.1. The system is sized to meet the load requirements with the lowest cost and the highest reliability. Therefore, this algorithm is designed to follow the *LOLP*, and E_{dummy} limits at the minimum value of *LEC*. *PVA*, *NWT*, DG size, Battery capacity, and *LEC* were the optimization parameters in this algorithm. P_{Dgr}, SOC_{max}, and SOC_{min} are sized to ensure the load demand at the time of insufficient generation of RES. Applying this algorithm, the best site of the sites under study and the most economic WT for the selected site can be determined.

3.2 Optimization Strategies of HRES

The objective function of the optimization problem is the minimization of the aggregate system cost function, *TPV(X)*. This function incorporates capital cost *IC* (*X*), operation and maintenance cost *OMC(X)*, the replacement cost *RC(X)* and cost of the diesel generator (*DGc*), throughout the lifetime of the installed system.

© Springer International Publishing AG 2018

M. Abdelaziz Mohamed and A.M. Eltamaly, *Modeling and Simulation of Smart Grid Integrated with Hybrid Renewable Energy Systems*, Studies in Systems, Decision and Control 121, DOI 10.1007/978-3-319-64795-1_3

The objective function for optimally designing the HRES must be minimized as expressed in the following equation:

$$\min_X TPV(X) = \min_X \{IC(X) + OMC(X) + RC(X) + DGc\} \qquad (3.1)$$

where, X is the vector of sizing variables; $X = NWT, PVA, P_{Dgr}$, and E_{BR}.

The proposed optimization strategies are summarized in the following subsections.

3.2.1 Management Strategy

The following strategy describes the proposed management algorithm of the HRES.

If the power generated from RES exceeds the power required for the load demand, the surplus power will be used to charge the batteries until go up to its maximum level, $E_{B,max}$. The extra power of the batteries will be used to supply the dummy load, P_{dummy}. This logic behind that is summarized as the following:

If $P_W(t) > P_L(t)$ and $SOC < E_{B,max}$ then;

$$P_{BC}(t) = [(P_W(t) - P_L(t)) \, \eta_{inv} + P_{PV}(t)] \eta_{BC}, \quad \text{charging process} \qquad (3.2)$$

If $P_W(t) > P_L(t)$ and $SOC \geq E_{B,max}$, then;

$$P_{BC}(t) = 0 \qquad (3.3)$$

$$P_{dummy}(t) = [(P_W(t) - P_L(t)) + P_{PV}(t) \, \eta_{inv}] \qquad (3.4)$$

If $P_W(t) < P_L(t)$, $[P_W(t) + (P_{PV}(t) \, \eta_{inv})] > P_L(t)$ and $SOC < E_{B,max}$ then;

$$P_{BC}(t) = \left[P_{PV}(t) - \frac{(P_L(t) - P_W(t))}{\eta_{inv}} \right] \eta_{BC}, \quad \text{charging process} \qquad (3.5)$$

If $P_W(t) < P_L(t)$, $[P_W(t) + (P_{PV}(t) \, \eta_{inv})] > P_L(t)$ and $SOC \geq E_{B,max}$ then;

$$P_{BC}(t) = 0 \qquad (3.6)$$

$$P_{dummy}(t) = [(P_W(t) + P_{PV}(t) \, \eta_{inv}) - P_L(t)] \qquad (3.7)$$

If the power required for the load demand exceeds the power generated from RES, the battery will be used to ensure the load demand until decreased to its minimum level, $E_{B,min}$. If still, there is a deficit power; the DG will be used to compensate the deficit load demand. This logic is prescribed in the following equations:

If $[P_W(t) + (P_{PV}(t)\,\eta_{inv})] < P_L(t)$ and $SOC > E_{B,min}$ then;

$$P_{BD}(t) = \frac{[P_L(t) - P_W(t) - (P_{PV}(t)\eta_{inv})]}{\eta_{inv}\eta_{BD}}, \quad \text{discharging process} \quad (3.8)$$

If $[P_W(t) + (P_{PV}(t)\,\eta_{inv})] < P_L(t)$ and $SOC \leq E_{B,min}$ then;

$$P_{BD}(t) = 0 \quad (3.9)$$

$$P_{Dg}(t) = [P_L(t) - P_W(t) - P_{PV}(t)]\mu_D \quad (3.10)$$

If $P_{Dg}(t) > P_{Dgr}(t)$ then;

$$LOLP = LOLP\,(t) + 1 \quad (3.11)$$

where, P_{BC} and P_{BD} are the battery charging and discharging power respectively, η_{inv} is the inverter efficiency (η_{inv} has been taken in this book as 95% in both directions) [83]. P_{Dg} is the power required from DG and P_{Dgr} is the rated power of DG. P_{Dgr} value estimation is based on the energy balance calculations which ensure feeding the load demand from HRES components contributions with the permitted value of $LOLP$. The optimization part selects the best value of P_{Dgr} which ensures feeding the load demand with the minimum LEC.

3.2.2 Optimization Strategy

The objective of the proposed optimization algorithm is to maximize the system reliability, which has been expressed in this book using the concept of $LOLP$. Another objective is to minimize E_{dummy} and allow its exploitation to feed dummy loads. Finally, the main objective is to achieve a minimum cost of the generated energy, which has been expressed in terms of LEC.

This section describes a proposed optimization algorithm that has been designed to follow the intended values of $LOLP$ and E_{dummy} to ensure the load demand with a minimum value of LEC. In this algorithm, the allowable value of $LOLP$ has been taken as 5% and the allowable value of E_{dummy} has been taken as 4% of the annual load demand, LAE.

The proposed algorithm ensures that the total energy generated from HRES satisfies the load requirements; otherwise, the size of the wind energy system or/and PV system must be increased by a certain value. Vice versa, if the total energy generated from HRES is greater than the load requirements, the size of the wind energy system or/and PV system must be reduced by a certain value. The cycle is repeated for each year until the generated energy just satisfies the load requirements within the specified constraints. The logic behind that can be summarized by the following steps:

- If $E_{dummy} \leq 0.0$, then increase NWT and PVA.
- If $E_{dummy} > 0.04 \times LAE$ then reduce NWT and PVA.
- If $LOLP > 0.05$ then increase NWT and PVA.

If $0.0 \leq E_{dummy} < 0.04 \times LAE$ and $0.0 < LOLP < 0.05$, then the optimum size of each component of the HRES can be obtained. The next step after determining the size of each component is to calculate the cost of energy.

The proposed algorithm uses one year of hourly data of wind speed, solar radiation, and ambient temperature for the sites under study. Also, the data for ten WT types from different manufacturers are used.

In the proposed algorithm, the penetration ratio (the ratio of wind generation to the total renewable generation) is changed with certain increments (10% in this study) to meet the load requirements of the sites under study. Based on the minimum LEC, the optimum penetration can be determined.

3.3 New Proposed Simulation Program

From the above-described strategies, a new proposed simulation program (NPSP) has been developed to carry out the sizing and optimization process. This program enables the optimization and simulation of the proposed HRES by conducting energy balance calculations for each hour during the year. The NPSP can determine the optimum size of each component of the HRES depending on the minimum cost of the generated energy, LEC. In addition, the program can select the best site from the available sites and the best WT for this site.

The NPSP has been written using MATLAB software. This program is flexible to change the optimization parameters and constraint limits (i.e. $LOLP$, E_{dummy} limits), unlike the software available on the market. The algorithm used in the NPSP is illustrated in the flowchart shown in Fig. 3.1.

To run the NPSP, the accompanying information must be accessible:

- The optimum design values, $LOLP$ and E_{dummy}.
- The geographic data of the sites under study and the meteorological data of wind speed, solar radiation, and temperature at these sites.
- Specification of WT, PV modules, inverter, batteries, and DG.
- The load power data.
- Technical and economic data of system components, lifetime, interest and inflation rates.

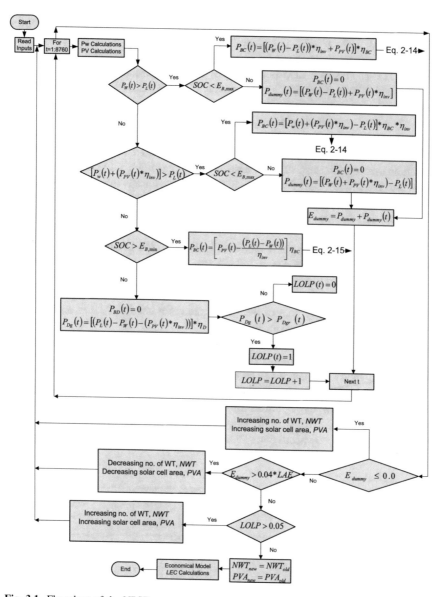

Fig. 3.1 Flowchart of the NPSP

3.4 Simulation Results and Discussion

The above algorithm has been applied for sizing and optimization of the proposed system to supply a load at five sites located in Saudi Arabia. The real data of the load shown in Fig. 2.4 has been used as a case study.

The simulation results showed that the lowest cost for kWh stands for Yanbu site and ITP-1 WT which is 33.650 (Cents/kWh), and using P_{Dgr} equal 44% of the peak load. So, we can say that the best site for HRES installation is Yanbu and the best WT is ITP-1. This WT also is the best WT for Qaisumah and Dhalm sites with 34.063 and 35.264 (Cents/kWh) respectively. This WT does not give the minimum *LEC* for the other two sites (Dhahran and Riyadh), where the minimum *LEC* for these two sites was achieved using Enercon-2 and Ecotecnia-2 WT with 36.879 and 40.812 (Cents/kWh) respectively. So, it is recommended to use ITP-1 WT in Yanbu, Qaisumah, and Dhalm sites and to use Enercon-2 and Ecotecnia-2 WT in Dhahran and Riyadh, respectively. Also, Yanbu, Qaisumah, and Dhalm are the best sites and the highest cost for energy is associated with Riyadh. Furthermore, it is clear that the cost of kWh generated in Riyadh is more than its value if we install the HRES in Dhahran, Dhalm, Yanbu, or Qaisumah. So it is not recommended to install HRES in Riyadh.

The average wind speed of Yanbu site, Qaisumah site, Dhalm site, Dhahran site, and Riyadh site are 4.9284 m/s, 4.7635, 4.589, 4.325, and 3.6979 m/s, respectively at 10 m above ground level. This gives an indication of the inverse relation between the average wind speed and the cost of the kWh of the site under study. Figures 3.2 and 3.3 show the hourly data of the wind speed at a height of 10 m and the tilted insolation for Yanbu site, respectively.

The optimization part is the important task of the NPSP, which is based on the iterative optimization technique. This part is intended to satisfy the system *LOLP*, E_{dummy}, and then calculate *LEC* to determine the optimal size of HRES which ensures the load demand. Figures 3.4, 3.5, 3.6, 3.7, 3.8, 3.9, 3.10, 3.11, 3.12, 3.13, 3.14, and 3.15 show the optimization iteration performances until achieving the optimum size of the HRES. Figures 3.4 and 3.15 show the first and the last optimization iterations, respectively. These figures show the performances of the

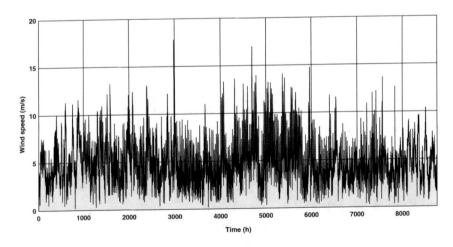

Fig. 3.2 Hourly wind speed for Yanbu site

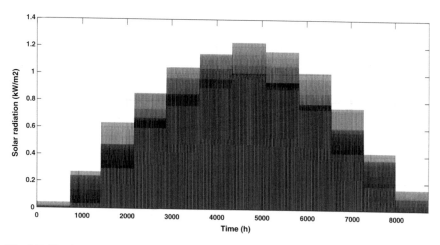

Fig. 3.3 Hourly tilted solar radiation for Yanbu site

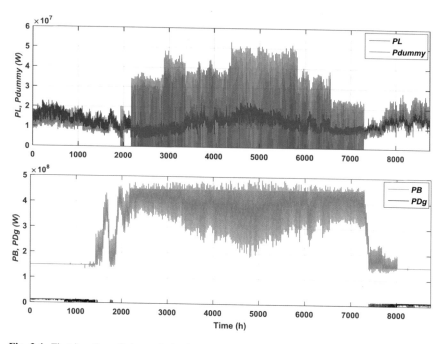

Fig. 3.4 First iteration of size optimization at Yanbu site

optimization process for the optimum case (Yanbu site, ITP-1 WT) at 50% penetration ratio. The iteration numbers vary depending on the optimization parameters and constraints. These figures describe the variation of the load power, PL, dummy load power, P_{dummy}, DG power, P_{Dg}, and the battery accumulated power, PB, with time until getting the optimum case.

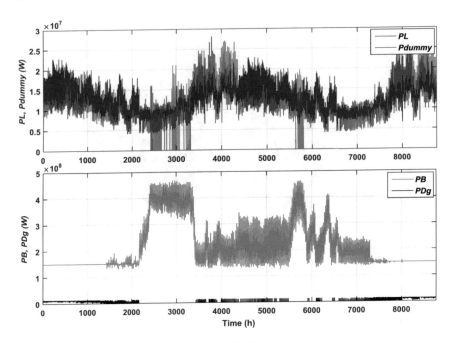

Fig. 3.5 Second iteration of size optimization at Yanbu site

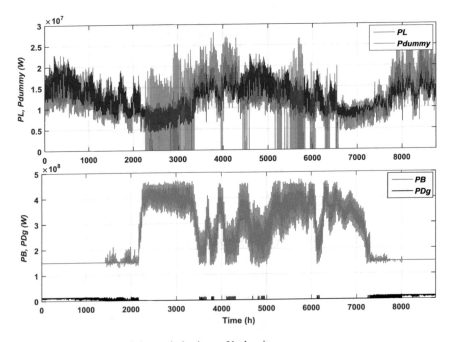

Fig. 3.6 Third iteration of size optimization at Yanbu site

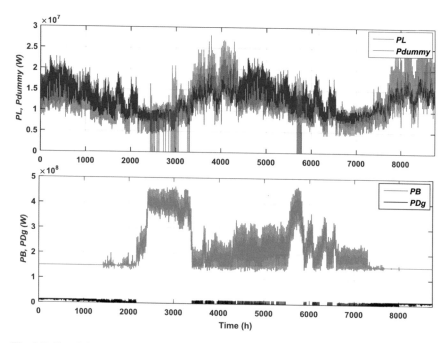

Fig. 3.7 Fourth iteration of size optimization at Yanbu site

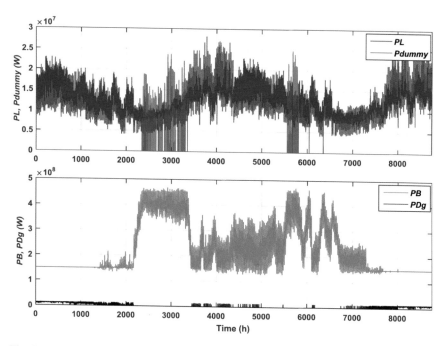

Fig. 3.8 Fifth iteration of size optimization at Yanbu site

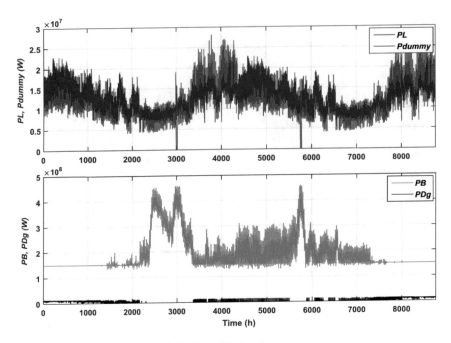

Fig. 3.9 Sixth iteration of size optimization at Yanbu site

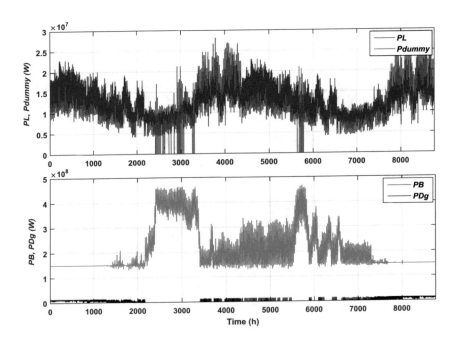

Fig. 3.10 Seventh iteration of size optimization at Yanbu site

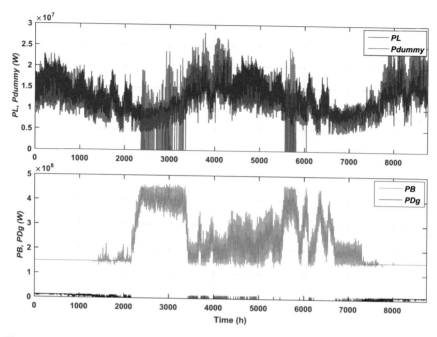

Fig. 3.11 Eighth iteration of size optimization at Yanbu site

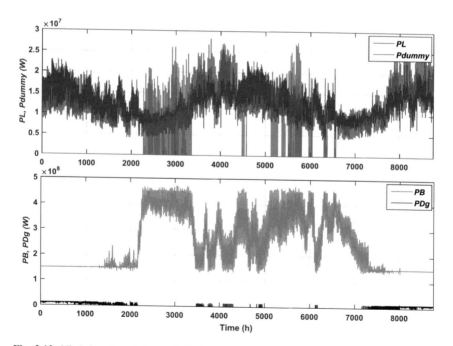

Fig. 3.12 Ninth iteration of size optimization at Yanbu site

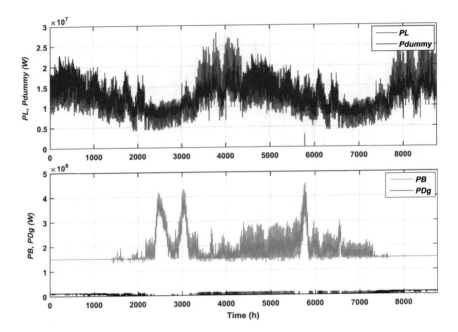

Fig. 3.13 Tenth iteration of size optimization at Yanbu site

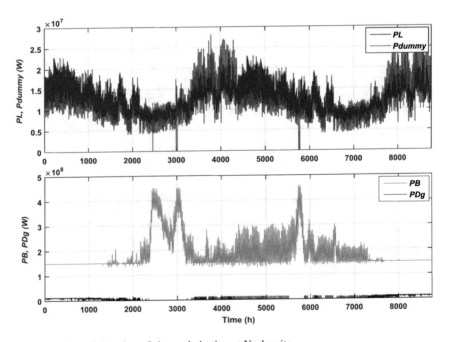

Fig. 3.14 Eleventh iteration of size optimization at Yanbu site

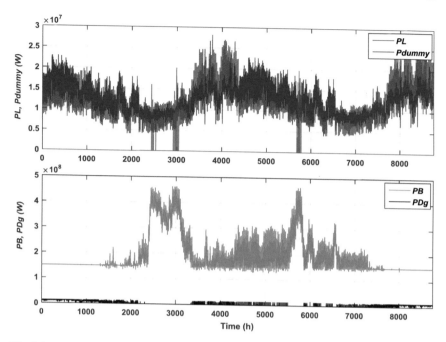

Fig. 3.15 Last iteration of size optimization at Yanbu site

Figure 3.16 shows the hourly variation of the HRES components at the optimum case. The HRES components which appear in this figure are load power (PL), power generated by the PV array and WT ($P_{PV} + P_W$), charging or discharging power of battery (P_{BC}/P_{BD}), DG power (P_{Dg}) and the dummy load power (P_{dummy}). According to this figure, the DG operates during the hours of insufficient generation of RES and inability of the batteries to supply the unmet load demand. Meanwhile, it is observed as expected that the dummy load absorbs the surplus renewable generation that exceeds the load demand and the battery demand.

Figure 3.17 shows a selected 24 h simulation results at the optimum case. This figure shows the variation of the load demand (PL), the power generated by both the PV array and WT ($P_{PV} + P_W$), and the power generated by the DG (P_{Dg}) on a certain day of the year as an example for better understanding of the logic behind the proposed algorithm.

Figure 3.18 shows the relation between the penetration ratio and LEC for Yanbu site and the ten WT under study (WT1–WT10). It is observed from this figure that the minimum value of LEC (i.e., the optimal value) is obtained at a 50% penetration ratio using WT5 at Yanbu site. The scale and shape parameters for this situation were 5.78 and 1.97 respectively.

Figure 3.19 shows the variation of LEC with the penetration ratio at the optimum case using a fourth-degree curve fitting. At the optimum case, the wind and PV contributions at 50% penetration are ($NWT = 67$ WT, $PVA = 10^5$ m^2).

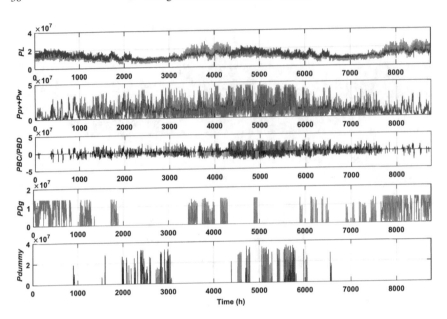

Fig. 3.16 Hourly simulation results of the optimum case

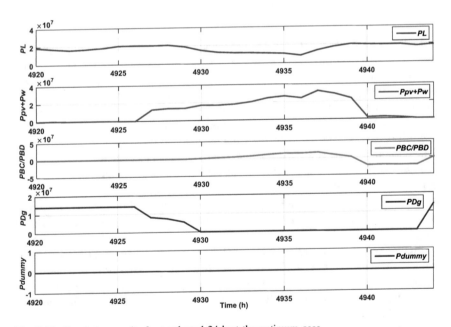

Fig. 3.17 Simulation results for a selected 24 h at the optimum case

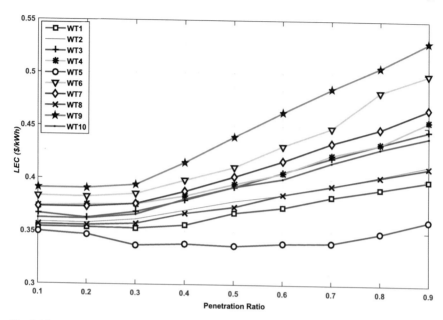

Fig. 3.18 Variation of the *LEC* with the penetration ratio for Yanbu site and ten WT under study

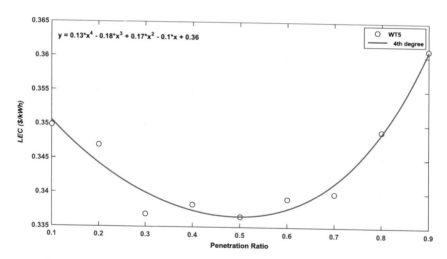

Fig. 3.19 Variation of the *LEC* with the penetration ratio for WT5

3.5 Conclusions

This chapter presents a new iterative optimization algorithm to solve the problem of sizing of HRES including PV, wind, diesel, and battery. The main objective of the proposed algorithm is to optimize the system to meet the load demand with minimum cost and highest reliability. The simulation results proved the ability of the proposed algorithm to determine the optimum solution, and the ability to change the optimization parameters and constraints. Furthermore, this algorithm can be applied on any site has it's meteorological data, and for any WT has its manufactures data. The results identified that Yanbu site is the best site and WT5 is the best one for this site. In addition, the best contribution from RES is 50% from wind energy system and 50% from PV energy system using 67 WT and 10^5 m^2 PVA with the cost of 33.65 Cents/kWh.

Chapter 4
A Novel Smart Grid Application for Optimal Sizing of Hybrid Renewable Energy Systems

4.1 Introduction

Utilization of various RES with storage and backup units to form HRES can give more economic and reliable source of energy [84]. But, due to the non-linear response of system components and the random nature of the RES and load, the smart grid is utilized to suit and incorporate these units in order to move the power around the system as efficiently and economically as possible [85, 86]. The HRES integrated into smart grid needs an accurate and optimum design and sizing to minimize the cost of generated energy and to maximize the system reliability.

In this chapter, a novel intelligent algorithm based on smart grid applications is introduced to determine the optimum size of stand-alone HRES so as to meet the load requirements with the minimum cost and the highest reliability. Also, load shifting-based load priority is presented as one of smart grid applications by dividing the load into two fractions, high priority load (HPL) and low priority load (LPL). HPL must be supplied whatever the generation conditions while LPL can be supplied from the surplus generation time of RES. Demand profile improvement has been carried out by shifting the LPL from low generation to high generation time.

4.2 Configuration of the Proposed Hybrid Renewable Energy System

The schematic diagram of the proposed stand-alone hybrid PV/wind/battery/diesel energy system is shown in Fig. 4.1. As shown in this figure, WT is connected to AC-bus, PV array is connected to DC-bus, and battery charger is connected to the DC-bus to charge the battery bank from the respective WT and PV array through a

© Springer International Publishing AG 2018
M. Abdelaziz Mohamed and A.M. Eltamaly, *Modeling and Simulation of Smart Grid Integrated with Hybrid Renewable Energy Systems*, Studies in Systems, Decision and Control 121, DOI 10.1007/978-3-319-64795-1_4

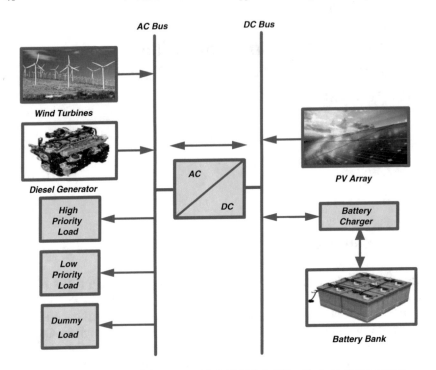

Fig. 4.1 Schematic diagram of the proposed hybrid PV/wind/diesel/battery energy system

bi-directional AC/DC converter. The DG is connected to AC-bus as a conventional source of energy. Finally, a group of loads, HPL, LPL, and dummy load are connected to AC-bus to consider the load demand for the HRES.

4.3   Proposed Power Management Strategy for HRES

The following operational algorithm is proposed for power management of the HRES:

The HPL is supplied primarily from WT and afterward PV array, respectively. In the case of the generated power from the RES surpasses the required power for the HPL (P_{LHP}), the excess power will be used to charge the batteries up to its maximum level, $E_{B,max}$. The abundance power above $E_{B,max}$ will be used to supply the LPL (P_{LLP}). If the power exceeds the LPL demand; the surplus power will be used to supply the dummy load, P_{dummy}. If there is an unmet LPL demand; it will be shifted to the time of surplus generation. This logic is condensed in the following points:

4.3.1 Battery Charge Mode

- If $P_W(t) > P_{LHP}(t)$ and $SOC < E_{B,\max}$, then;

$$P_{BC}(t) = [(P_W(t) - P_{LHP}(t))\eta_{inv} + P_{PV}(t)]\eta_{BC} \qquad (4.1)$$

- If $P_W(t) < P_{LHP}(t)$, $[P_W(t) + (P_{PV}(t)\eta_{inv})] > P_{LHP}(t)$ and $SOC < E_{B,\max}$ then;

$$P_{BC}(t) = \left[P_{PV}(t) - \frac{(P_{LHP}(t) - P_W(t))}{\eta_{inv}}\right]\eta_{BC} \qquad (4.2)$$

In all cases of battery charge mode the amassed value of the unmet LPL, P_{LLP_sum}, and SOC of the battery can be calculated from the following mathematical statements:

$$P_{LLP\_sum} = P_{LLP\_sum} + P_{LLP}(t) \qquad (4.3)$$

$$E_B(t+1) = E_B(t)(1 - \sigma) + P_{BC}(t) \qquad (4.4)$$

4.3.2 Feeding the Low Priority Load

- If $P_W(t) > P_{LHP}(t)$ and $SOC \geq E_{B,\max}$ then;

$$P_{L\_low}(t) = [(P_W(t) - P_{LHP}(t)) + P_{PV}(t)\eta_{inv}] \qquad (4.5)$$

- If $P_W(t) < P_{LHP}(t)$, $[P_W(t) + (P_{PV}(t)\eta_{inv})] > P_{LHP}(t)$ and $SOC \geq E_{B,\max}$ then;

$$P_{L\_low}(t) = [(P_W(t) + (P_{PV}(t)\eta_{inv})) - P_{LHP}(t)] \qquad (4.6)$$

In all cases of feeding LPL, the accumulation of the unmet LPL and SOC of the battery can be computed from the following equations:

$$P_{LLP\_sum} = P_{LLP\_sum} + P_{LLP}(t) - P_{L\_low}(t) \qquad (4.7)$$

$$E_B(t+1) = E_B(t)(1 - \sigma) \qquad (4.8)$$

4.3.3   Feeding the Dummy Load

Dummy load can be used to absorb the surplus renewable generation that exceeds the LPL demand, and battery demand as shown in the following equations:

- If $P_{L\_low}(t) > P_{LLP\_sum}$

$$P_{dummy}(t) = P_W(t) + P_{PV}(t)\eta_{inv} - P_{LHP}(t) - P_{LLP\_sum} \qquad (4.9)$$

4.3.4   Battery Discharge Mode

If the RES couldn't meet the power required for the HPL, batteries will be utilized to cover the HPL demand until reductions to their minimum level, $E_{B,min}$. The unmet LPL will be shifted to the time of surplus generation. This logic is summarized in the accompanying equations:

- If $[P_W(t) + (P_{PV}(t)\eta_{inv})] < P_{LHP}(t)$ and $SOC > E_{B,min}$ then;

$$P_{BD}(t) = \frac{[P_{LHP}(t) - P_W(t) - (P_{PV}(t)\eta_{inv})]}{\eta_{inv}\eta_{BD}} \qquad (4.10)$$

$$E_B(t+1) = E_B(t)(1 - \sigma) - P_{BD}(t) \qquad (4.11)$$

$$P_{LLP\_sum} = P_{LLP\_sum} + P_{LLP}(t) \qquad (4.12)$$

where, P_{L_low} is the accessible power used to supply the LPL.

4.3.5   Diesel Generator Operation

If the produced power from RES and the battery are not adequate to supply HPL, the deficit power in HPL will be compensated using the DG as expressed in the following equations:

- If $[P_W(t) + (P_{PV}(t)\eta_{inv})] < P_{LHP}(t)$ and $SOC \leq E_{B,min}$ then;

$$P_{Dg}(t) = [(P_{LHP}(t) - P_W(t) - (P_{PV}(t)\eta_{inv}))] \qquad (4.13)$$

4.4 Problem Statement

The aim of this chapter is to introduce an algorithm based on smart grid applications for sizing the HRES to supply the load demand while considering the minimum cost and satisfying a defined reliability index.

Demand profile improvement as one of the essential smart grid applications has been covered in this chapter. Demand profile improvement helps in smoothing the demand profile, and/or reducing the peak demand or the total energy demand. This will diminish the overall plant and capital cost prerequisites, the cost of the generated energy, and furthermore will increase the system reliability. Demand profile improvement is carried out in this chapter by shifting the LPL from low generation to high generation time of RES.

In the above algorithm, cost estimation is based on *LEC* and the reliability of the HRES is produced in view of the concept of *LOLP*. Cost estimation and reliability assessment of the HRES are detailed in the following subsections.

4.4.1 Reliability Assessment

In the case of the generation from HRES components is insufficient to sustain the HPL and/or LPL, then the load will not be supplied and the system will lose its reliability. On account of unmet HPL, the *LOLP* counter will increment when this circumstance happens. The following equations clarify the system operation amid this condition:

- If $P_W(t) + P_{PV}(t)\eta_{inv} + P_{Dg}(t) + P_{BD}(t) < P_{HPL}(t)$ then;

$$LOLP\_HP = LOLP\_HP + 1 \tag{4.14}$$

where, *LOLP_HP* is the counter for loss of load probability of HPL.

For this situation, the counter P_{LLP_sum} will increase as given in the mathematical statement (4.12).

4.5 Formulation of the Optimum Sizing Problem

Size estimation of the hybrid PV/wind/battery/diesel energy system is formulated as an optimization problem and the objective function is formulated corresponding to system constraints and performances. The discussion on the objective function and the constraints is detailed in the following subsections.

4.5.1 Objective Function

The objective function of the optimization problem is to minimize the overall system cost $TPV(X)$ as expressed in Eq. (3.1).

4.5.2 Design Constraints

To solve the optimization problem, a set of constraints that must be satisfied with any feasible solution throughout the system operations are as follows:

- At any time, the *SOC* of the battery bank should satisfy the following constraints:

$$E_{B,\min} \leq E_B(t) \leq E_{B,\max} \tag{4.15}$$

$$E_B(t+1) = E_B(t)(1 - \sigma) \tag{4.16}$$

- At any time, the hourly power generated by diesel generator, P_{Dg} should be less than or equal to the diesel generator rated power, P_{Dgr} as shown in the following expression:

$$P_{Dg}(t) \leq P_{Dgr} \tag{4.17}$$

- *LOLP* of the system should be less than allowable *LOLP* reliability index as shown in the following expressions:

$$LOLP\_HP < LOLP\_HP_{index} \tag{4.18}$$

$$P_{LL\_sum} < P_{LL\_sum_{index}} \tag{4.19}$$

where, $LOLP\_HP_{index}$ and $P_{LL\_sum_{index}}$ are the designed values of the counters $LOLP\_HP$ and P_{LL_sum}, respectively which are specified by the user.

- Dummy energy, E_{dummy} should satisfy the following constraint:

$$E_{dummy_{min}} < E_{dummy} < E_{dummy_{max}} \tag{4.20}$$

where, $E_{dummy_{min}}$ and $E_{dummy_{max}}$ are the minimum and the maximum allowable values of the dummy energy and specified by the user.

4.5.3 A Proposed Optimization Algorithm for HRES

In this section, a proposed optimization algorithm has been designed to follow the intended values of $LOLP$ and E_{dummy} of the HRES to fulfill an aggregate load demand with minimum LEC. In this algorithm, the value of $LOLP\_HP_{index}$ has been considered to be 4% and $P_{LL\_sum_{index}}$ has been taken by (8 days of average LPL, $8 \times PL_{ave\_low}$). $E_{dummy_{min}}$ and $E_{dummy_{max}}$ have been considered to be 0%, 4% of LAE, respectively.

The proposed algorithm ensures that the total energy generated from HRES must satisfy the load requirements; otherwise, NWT and/or PVA ought to be increased by specific values and vice versa. The cycle begins again until satisfying the demand requirements, accomplish the objective function and constraints. Along these lines, if $(0 < E_{dummy} < 0.04 \times LAE)$, $(P_{LLP\_sum} < 8 \times 24 \times PL_{ave\_low})$ and $(LOLP\_HP < 0.04)$, then the optimum size of the HERS components can be obtained. The following step after deciding the optimum size of HERS component is to compute the LEC.

4.6 New Proposed Program Based Smart Grid

From the above description of different component models and the power management, a new proposed program based-smart grid (NPPBSG) has been developed to perform the sizing and optimization process of HRES. This program allows the design and simulation of the proposed HRES by conducting the energy balance calculations for each hour during the year based on the smart grid applications. NPPBSG has been written in MATLAB software in a flexible fashion and allows the variation of the optimization parameters and constraints.

Hourly data of wind speed, solar radiation, temperature, and load power for the sites under study have been used through this program. The penetration ratio has been changed with increment 10% to meet the load requirements of the sites under study. Utilizing NPPBSG, the best site and the most economic WT for the selected site can be determined.

The logic used in the NPPBSG is illustrated in the flowchart shown in Fig. 4.2 which summarize the above discussed smart grid logic. The required data for NPPBSG are listed in the following points:

- The optimum design values; $LOLP\_HP_{index}$, $P_{LL\_sum_{index}}$, $E_{dummy_{min}}$ and $E_{dummy_{max}}$.
- The geographic data of the sites under study and meteorological data of wind speed, solar radiation, and temperature in these sites.
- Specification of WT, PV modules, inverter, batteries, and diesel generator.
- The load power data, HPL, and LPL.
- Technical and economic data of system components, lifetime, interest rate and inflation rate.

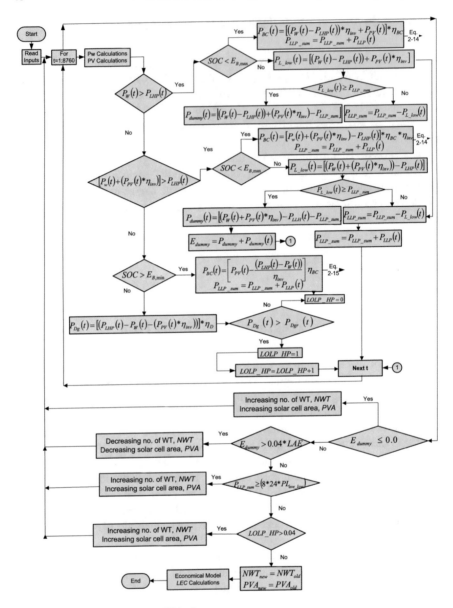

Fig. 4.2 Flowchart of the NPPBSG logic

4.7   Simulation Results and Discussions

The proposed algorithm has been applied to carry out sizing and optimization of a stand-alone HRES in order to supply a certain load at five sites located in Saudi Arabia using smart grid applications.

The real data of the load shown in Fig. (2.4) has been used as a case study. This load has been divided into two priority-based load types, HPL, and LPL. The HPL has been considered to be 75% of the total load and LPL has been considered to be 25% of the total load.

The simulation results identified Yanbu site as the best site among the sites under study and ITP-1 WT as the best WT for this site.

The optimization part is an essential part in the NPPBSG which uses the iterative optimization method. This part is designed to follow the system design values, P_{LLP_sum}, $LOLP\_HP$, and E_{dummy} in objective to determine the optimal size of the system to meet the annual load demand with minimum cost.

The first and last optimization iterations until get the optimum design are shown in Figs. 4.3 and 4.4, respectively. The hourly variation of load power, PL and dummy power, P_{dummy} are shown in Figs. 4.3a and 4.4a. The DG power, P_{Dg} and the battery accumulated power, PB are shown in Figs. 4.3b and 4.4b. The accumulated unmet power of LPL, P_{LLP_sum} is shown in Figs. 4.3c and 4.4c. As shown in Fig. 4.3a, E_{dummy} didn't satisfy the optimum value, therefore, applying NPPBSG the optimum value has been acquired as is clear from Fig. 4.4a. The NPPBSG guarantees to supply the LPL demand as the year progressed, and also, permit low rate of unmet LPL demand to be shifted to the following year, as shown in Figs. 4.3c and 4.4c. Moreover, the NPPBSG seeks to distribute the LPL demand through the year using load shifting, to ensure the load not to be concentrated and to reduce the peak load demand, as shown in Figs. 4.3c and 4.4c.

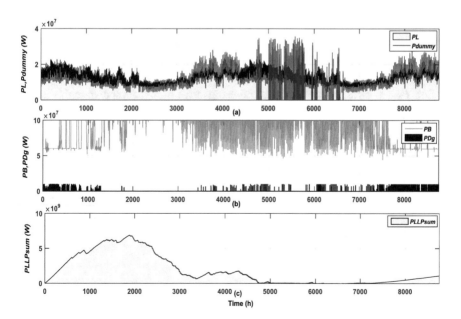

Fig. 4.3 The first optimization iteration of the optimum case

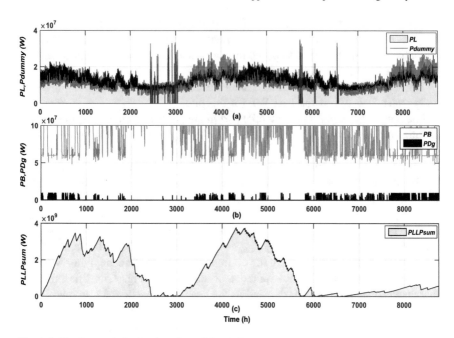

Fig. 4.4 The last optimization iteration of the optimum case

Figure 4.5 shows the performance of the system components with time for the optimum case. This figure describes the response of the HRES components with varying the meteorological data and load profile with 25% load shifting.

Figure 4.6 describes the sensitivity of the NPPBSG if different load shifting percentages are applied. The figure shows the relation between load shifting ratio [LPL energy/total Energy (*ELPL/LAE*)] and the DG size as a ratio of the DG rated value, P_{Dgr} at 50% penetration ratio and using curve fitting. As shown in this figure, DG size decreases with the increasing of load shifting, which in turn reduces the system capital cost, CO_2 emission, and *LEC*.

The variation of *LEC* with the penetration ratio, with and without load shifting has been described in Fig. 4.7. It is observed in this figure, that the cost of the generated energy, *LEC* has been reduced with load shifting.

Figure 4.8 shows a comparison between no load shifting performance and LPL shifting. As shown in this figure, all cost calculations, *LEC* and the battery capacity have been reduced with load shifting.

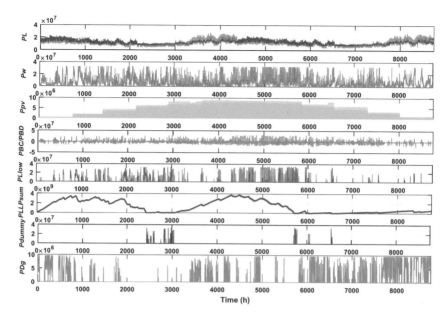

Fig. 4.5 Performance of the HRES components with time

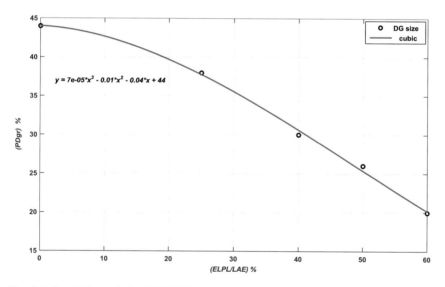

Fig. 4.6 Sensitivity analysis of NPPBSG for different load shifting percentages

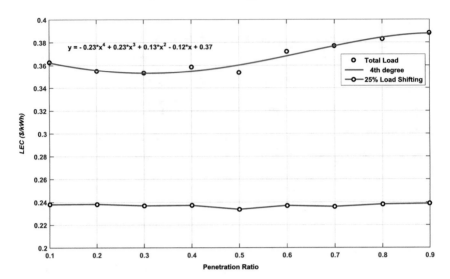

Fig. 4.7 The relation between *LEC* and penetration ratio with and without load shifting

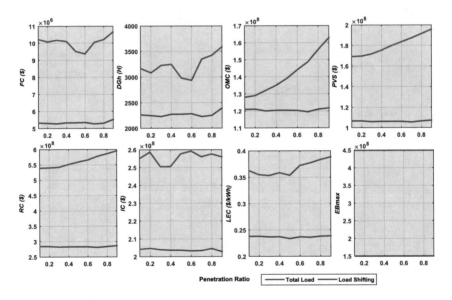

Fig. 4.8 A comparison between no load shifting performance and LPL shifting

4.8 Conclusions

In this chapter, a proposed techno-economic algorithm and a proposed program based on smart grid applications have been introduced. The main objective of this algorithm is to optimally design a hybrid PV/wind/diesel/battery energy system in order to meet the load requirements with minimum cost and highest reliability.

Smart grid applications have been applied in this chapter by dividing the load into high and low priorities. HPL must be supplied at any time from PV/wind systems, battery, and/or diesel generator. But, LPL can be supplied only in case of the generation from PV/wind is greater than HPL and the battery is fully charged, otherwise, low priority load will be shifted to next hour.

Hybrid AC/DC-buses coupling has been used for the proposed model to increase the system efficiency and reliability. Furthermore, the data for five sites in Saudi Arabia and ten WT from different manufacturers have been used as a case study and an accurate pairing between these sites and WT has been carried out. In addition, the sensitivity analysis has been carried out in this study to check and validate the response of the proposed algorithm under varying the operating conditions.

The simulation results confirmed the effectiveness of the proposed program for determining the optimum solution and assured the capability of this program to adapt to varying operating conditions. Moreover, the simulation results confirmed that utilizing load shifting can reduce the system size, reduces the cost of generated energy and increases the system reliability.

Chapter 5
A PSO-Based Smart Grid Application for Optimum Sizing of Hybrid Renewable Energy Systems

5.1 Introduction

One of the most important issues in the recent studies is to optimally size the HRES components to meet all load requirements with possible minimum cost and highest reliability. In view of the complexity of optimization of the HRES, it was imperative to discover effective optimization methods ready to get good optimization results especially, for the complex optimization problems. Particle swarm optimization (PSO) was recommended as a standout amongst the most valuable and promising methods for optimizing the HRES because of using the global optimum to locate the best solution [87]. PSO algorithm is designed based on swarm intelligence and used to handle the complex optimization problems [88].

In this chapter, a novel intelligent PSO algorithm based on smart grid applications is introduced to determine the optimum size of the hybrid system shown in Fig. 4.1. The main objective of this algorithm is to optimize the system size so as to meet the load requirements with the minimum cost and the highest reliability.

Load shifting-based load priority is presented in this chapter as one of smart grid applications by shifting the LPL from deficit generation time to surplus generation time of HRES.

A comparison between the results obtained from PSO algorithm and those from the iterative optimization techniques (IOT) is introduced. Moreover, parallel implementation of PSO (PIPSO) is a new proposed method utilized in this book to distribute the evaluation of the fitness function and constraints among the ready-made processors or cores, and to speed up the optimization process. Furthermore, a comparison between utilizing PIPSO and utilizing a serial implementation of PSO (SIPSO) is presented.

© Springer International Publishing AG 2018
M. Abdelaziz Mohamed and A.M. Eltamaly, *Modeling and Simulation of Smart Grid Integrated with Hybrid Renewable Energy Systems*, Studies in Systems, Decision and Control 121, DOI 10.1007/978-3-319-64795-1_5

5.2 Implementation of Particle Swarm Optimization Algorithm

PSO is a multi-agent parallel search optimization technique, which was presented in 1995 by Kennedy and Eberhart [89]. PSO is an evolutionary technique which is inspired by the social behavior of bird flocking, fish schooling and swarm application [90, 91]. Each particle in the PSO algorithm represents a potential solution; these solutions are assessed by the optimization objective function to determine their fitness.

In order to move to the optimum solution, particles move around in a multidimensional search space. The best experience for each particle is stored in the particle memory and called local best particle (*pbest$_i$*) and the best global obtained among all particles is called as a global best particle (*gbest*). During flight the current position (x_i) and velocity (v_i) of each particle (i) is adapted according to its own experience and the experience of neighboring particles as described by the following equations [89]:

$$v_i^{(g+1)} = \omega v_i^{(g)} + c_1 a_1 \left(pbest_i - x_i^{(g)} \right) + c_2 a_2 \left(gbest - x_i^{(g)} \right) \qquad (5.1)$$

$$x_i^{(g+1)} = x_i^{(g)} + v_i^{(g+1)} \qquad (5.2)$$

where, g is the counter of generations, and ω is the inertia weight factor (i.e. inertial velocity of the particles) in a range of [0.5, 1] [92]. c_1 and c_2 are positive acceleration constants in a range of [0, 4], designated as self-confidence factor and swarm confidence factor, respectively [92]. These factors provide an insight from a sociological stand point. c_1 has a contribution towards the self-experience of particles. c_2 has contribution towards the motion of the particles in global direction. a_1 and a_2 are uniform randomly generated numbers in a range of [0, 1] [92].

Swarm size, the number of particles, ω, c_1 and c_2 are the main parameters of the PSO algorithm, which are initialized by the users, based on the problem being optimized.

To execute the proposed management and optimization procedures of the HRES, a new proposed program based-PSO (NPPBPSO) has been developed. NPPBPSO has been written using MATLAB software in a flexible fashion that is not available in the recent market software such as HySys, HOMER, iHOGA, iGRHYSO, HYBRIDS, RAPSIM, SOMES, HySim, IPSYS, ARES, and SOLSIM [93]. The NPPBPSO gives the possibility to change the penetration ratio with a certain increment (10% in this study) to decide the optimum contribution from RES (i.e. the best penetration ratio). This program can determine the optimum size of the HRES components for supplying the load demand with minimum *LEC* and within the specified limits of *LOLP* and *P$_{dummy}$*. Likewise, the program can choose the best site out of many available sites and select the most economic WT for this site. In this program, the value of *LOLP_HP$_{index}$* has been considered to be 4% and

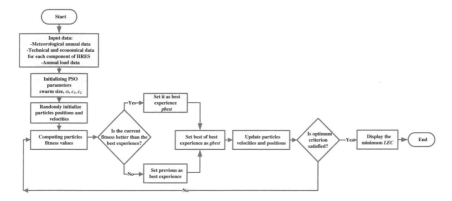

Fig. 5.1 The process of the PSO algorithm

$P_{LL\_sum\,index}$ has been taken by (8 days of average LPL, 8 $\times$ PL_{ave_low}). $E_{dummy\,min}$ and $E_{dummy\,max}$ have been considered to be 0%, 4% of *LAE*, respectively. The process of the PSO algorithm is shown in Fig. 5.1.

To run NPPBPSO the following information must be accessible:

- Initial values of PSO parameters, swarm size, the number of particles, ω, c_1 and c_2.
- The optimum design values; $LOLP\_HP_{index}$, $P_{LL\_sum\,index}$, $E_{dummy\,min}$ and $E_{dummy\,max}$.
- The geographic data of the sites under study and meteorological data of wind speed, solar radiation, and temperature at these sites.
- Specification of WT, PV modules, inverter, batteries, and diesel generator.
- The load power data, HPL, and LPL.
- Technical and economic data of system components, lifetime, interest rate and inflation rate.

5.3   Parallel Implementation of Particle Swarm Optimization

PSO is considers an effective tool for nonlinear optimization problems. However, PSO sometimes requires a lengthy computing time to find the optimal solution. Parallel implementation of particle swarm optimization (PIPSO) is a set of parallel algorithms, in which the population is divided into sub-populations that advance autonomously. Every sub-population is allocated in each processor included in parallel computing. PIPSO can automatically distribute the evaluation of the fitness function and constraints among the ready-made processors or cores. PIPSO is likely to be faster and time saver than the serial implementation of particle swarm optimization (SIPSO); when computation of the fitness function is time-consuming, or when there are many particles. Otherwise, the overhead of distributing the evaluation among the ready-made processors or cores can cause PIPSO to be slower than SIPSO [94]. In this study, PSO has been used to search for the optimal solution, and PIPSO has been used to overcome the shortcomings of PSO. To use PIPSO a license for parallel computing toolbox software and a parallel worker pool (parpool) must be available.

5.4 Results and Discussion

A PSO-based MATLAB algorithm has been created to determine the optimum size of a PV/WT/batteries/DG system in order to supply a certain load in different remote sites in Saudi Arabia.

As indicated in the literature and according to the nature of the problem under study; the suitable values for PSO parameters have been set to make the PSO faster, and exact. The population size has been set to be 20, a maximum number of iterations has been set to be 100, c_1 and c_2 have been chosen as 2, a_1 and a_2 have been picked as 0.02, and ω has been set as 0.7.

In addition, SOC of the battery bank and the DG rated power have been sized to meet the load demand in the time of deficit generation. HPL and LPL have been considered to be 75 and 25% of the aggregate load demand, respectively.

The optimization part of the NPPBPSO is the key part which is used to optimize the size of the HRES components to fulfill the load demand, with minimum cost while satisfying the optimization constraints. PVA, NWT, P_{Dgr}, and E_{BR} are the optimization parameters in NPPBPSO.

After initiating the PSO parameters, the PSO algorithm is applied to get the optimum case, the NPPBPSO results affirmed that the minimum cost of energy for the specified limits of $LOLP$ and E_{dummy} was in Yanbu site and the best WT for this site was ITP-1. Additionally, the best share from the RES was at 50% penetration ratio.

The outcomes acquired from NPPBPSO have been compared with those obtained from IOT. The comparison results are shown in Table 5.1. As seen in this table, the results obtained from the IOT and NPPBPSO are almost the same. The PSO idea relies on imposing various particles for searching the optimum solution, every particle represents a solution. In the next iteration, the solutions number is multiplied until it gets the optimum one. Imposing more particles in each iteration encourage coming to the optimum solution, and furthermore decreases the

Table 5.1 The comparison results of IOT and NPPBPSO

Tools	c	k	NWT	PVA	PSV	FC	OMC	RC	LEC
IOT	5.78	1.97	91	$3.80 * 10^4$	$1.80 * 10^8$	$7.18 * 10^6$	$1.2 * 10^8$	$2.9 * 10^8$	0.2417
NPPBPSO	5.72	1.95	90	$3.78 * 10^4$	$1.70 * 10^8$	$6.89 * 10^6$	$1.1 * 10^8$	$2.8 * 10^8$	0.2334

number of optimization iterations. The solution obtained from PSO is the one that fulfills all the optimization constraints and objective function, which makes it the exact solution. The IOT impose one solution for each iteration relying upon the experimentation strategy, which in turn may raise the number of iterations until getting the optimum solution. The solution got from IOT is the one that satisfies the optimization constraints and objective functions, yet this solution may be away from the optimum one. In this way, the most minimal cost acquired with the IOT is not the optimum solution but rather it will be the best plausibility from the accessible solutions. The main advantages of the PSO are its ability to solve the complex certifiable problems, high adaptability and ability to manage nonlinear, non-differentiable functions and functions with an expansive number of parameters. However, the IOT can't solve variant optimization problems attributable to poorly known objective functions and that have multi-constraints.

Figure 5.2 demonstrates the convergence process of the PSO algorithm during the minimization of the *LEC* for 4 autonomous runs. As illustrated in this figure, the optimum solution is acquired after around 30 iterations, and the 100 iterations are considered as a reasonable end measure. In addition, it can be noted that the optimum solution almost converges to the same optimum value (global minimum) for all runs.

Figure 5.3 shows the convergence process for one run of the IOT. As shown in this figure, the optimum solution is obtained after around 215 iterations and this solution may be away from the optimum one. It is also observed that the time taken to find the optimum sizing by using PSO is lower than that taken by using IOT. Therefore, the optimization utilizing PSO is quicker and more precise than utilizing IOT.

PIPSO is a granulated approach to speed up the optimization process, to activate PIPSO, parallel choice should set to true. When this condition is true, the NPPBPSO assesses the objective function of the optimization problem in parallel

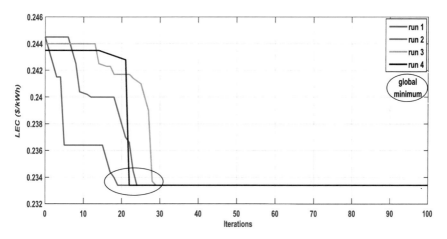

Fig. 5.2 The convergence process of the PSO algorithm for 4 autonomous runs

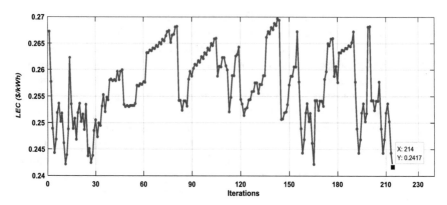

Fig. 5.3 The convergence process of the IOT

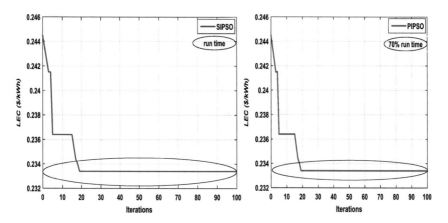

Fig. 5.4 The comparison results of SIPSO and PIPSO

population. Figure 5.4 shows how to speed up the optimization process by utilizing the PIPSO. Intel® Core™ i5-2410 M processor with clock speed: 2.30/2.90 Turbo GHz, 3rd level cache: 3 MB and front side bus: 1333 MHz has been used to run the optimization process. The optimization process has been carried out in a serial manner as appeared in the first part of Fig. 5.4 (SIPSO) and carried out in the second part of Fig. 5.4 using parpool (PIPSO). As clear from this figure, utilizing the PIPSO can save more time during the optimization process.

Figure 5.5 describes the sensitivity analysis of the proposed algorithm with the help of quadratic curve fitting. This figure shows the DG performance with relation to the rate of load shifting, at 50% penetration ratio. As shown in this figure, there is an opposite relation between load shifting rate (LPL energy (*LPLE*)/annual load energy (*LAE*)) and the DG capacity (P_{Dgr}). In other words, as the rate of load

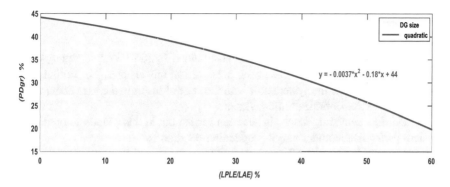

Fig. 5.5 Sensitivity analysis for DG performance with load shifting rate

shifting increases, it reduces the DG capacity, which in turn, reduces the whole system cost, and also, reduces the load peak value.

The NPPBPSO has been applied to design and optimize the HRES to ensure the load demands in two cases, one for feeding the full load demand and the other one for feeding the load demand with shifting the LPL. A comparison between these two cases is shown in Fig. 5.6. This figure shows that utilizing load shifting-based load priority reduces the whole system cost, *LEC* and reduces the size of the HRES components, DG capacity, and the battery capacity. Additionally, it lessens the aggregate operation hours of the DG through the system lifetime, and thus diminishes the CO_2 emission and environment contamination.

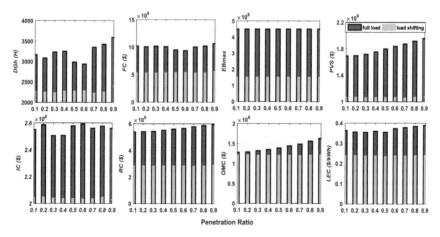

Fig. 5.6 A comparison between full load performance and load shifting performance of HRES with penetration ratio

5.5 Conclusions

A methodology for optimum sizing of stand-alone hybrid PV/wind/battery/diesel energy systems utilizing PSO has been presented in this chapter. The optimization goal was to minimize the system cost with the state of insuring the load demand and satisfying a set of optimization constraints.

In addition, sensitivity analysis has been carried out in this chapter to predict the system performance under varying operating conditions.

The simulation results affirmed that PSO is the promising optimization techniques due to its ability to reach the global optimum with relative simplicity and computational proficiency contrasted with the customary optimization techniques.

Finally, PIPSO has been utilized to speed up the optimization process, and the simulation results confirmed that it can save more time during the optimization process compared with utilizing SIPSO.

Chapter 6
General Conclusion

6.1 General Conclusion

In this book, a proposed system that includes PV, WT, DG as a conventional source of energy, and battery bank for energy storage to form a hybrid energy system has been presented. The system components have been connected through mixed AC/DC-buses in order to increase the system reliability and efficiency. This system is used to supply a certain load demand of a stand-alone system.

The main objective of this book is to optimize the size of the proposed system in order to supply the load with the minimum cost and the highest reliability. To achieve this objective, a proposed algorithm has been introduced to optimize the system size while satisfying the optimization constraints. The proposed algorithm has been designed to follow the reliability index of the system which has been represented in this book by using *LOLP* index, and to follow the cost of the generated energy which has been represented by using *LEC* concept. The decision variables included in the optimization process are *PVA*, *NWT*, battery capacity, and DG capacity.

Load shifting as one of the smart grid applications has been presented seeking for demand profile improvement, peak value reduction, CO_2 emission control, and reduction of the whole system cost. To achieve these targets, the load demand has been divided into two loads based-priority, HPL, and LPL. Load shifting has been carried out by shifting the LPL from the time of deficit generation to the time of surplus generation.

IOT and the PSO algorithm have been used as a part of this book to carry out the optimization process to find the optimum solution. The results obtained from IOT have been compared with those obtained from PSO algorithm. Also, PIPSO has been presented in this book as a new proposed technique, and a comparison between using PIPSO and SIPSO has been introduced.

The above algorithm has been implemented as a new program to carry out the optimization process using MATLAB software. This program is flexible enough to

© Springer International Publishing AG 2018   61
M. Abdelaziz Mohamed and A.M. Eltamaly, *Modeling and Simulation of Smart Grid Integrated with Hybrid Renewable Energy Systems*, Studies in Systems, Decision and Control 121, DOI 10.1007/978-3-319-64795-1_6

deal with the change of the operating conditions which is not available in the famous market available software.

The proposed algorithm has been applied on five stand-alone sites in Saudi Arabia and ten different types of WT as a case study. Real meteorological data of wind speed, solar radiation, and temperature of these sites have been used through this algorithm. Also, the load data and the WT specifications have been used.

The simulation results can be concluded in the following points:

- The IOT is an effective method to get the optimum solution of the optimization problem. But, the number of optimization iterations to get the optimum solution may be large, depending on the optimization constraints and the designed values of $LOLP$, P_{LLP_sum} and E_{dummy}. This, in turn, may consume more time to get the optimum solution.
- The solution obtained from the IOT within the specified range of the optimization constraints ($LOLP$, P_{LLP_sum}, and E_{dummy}) and satisfies the optimization constraints, but may be away from the optimum one (global minimum cost). The accurate optimum solution depends on the designed range of the optimization constraints. Whenever a narrow range whenever the most accurate solution but, in turn, will be a difficult task on the IOT.
- PSO is the promising optimization techniques due to its ability to reach the global optimum with relative simplicity and computational proficiency contrasted with the customary optimization techniques. PSO is generally robust in finding a global optimum solution, especially in multi-objective optimization problems, where the determination of the global optimum is a difficult task.
- The solution obtained from PSO is an accurate one, because PSO assuring satisfying the constraints and the objective function, which is the minimum cost in this study.
- It is also observed that the time taken to find the optimum solution by using PSO is lower than that taken by using IOT. Therefore, the optimization utilizing PSO is quicker and more precise than using IOT.
- PIPSO can be utilized to distribute and speed up the optimization process, which in turn, save more time compared with SIPSO.
- Load shifting as one of smart grid applications can be applied to provide a distributed load profile, reduce the entire system cost and reduce CO_2 emission.

Appendix A
Wind Speed for Sites Under Study

See Figs. A.1, A.2, A.3 and A.4.

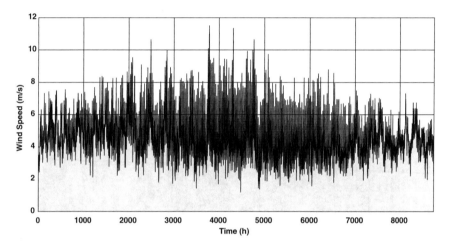

Fig. A.1 Hourly wind speed for Dhahran site

© Springer International Publishing AG 2018

M. Abdelaziz Mohamed and A.M. Eltamaly, *Modeling and Simulation of Smart
Grid Integrated with Hybrid Renewable Energy Systems*, Studies in Systems,
Decision and Control 121, DOI 10.1007/978-3-319-64795-1

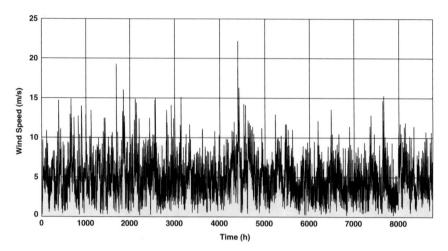

Fig. A.2 Hourly wind speed for Dhalm site

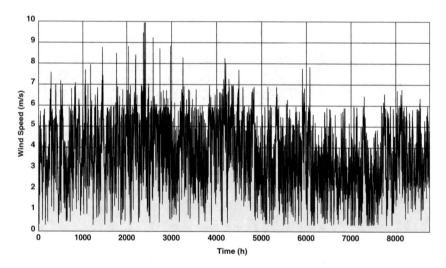

Fig. A.3 Hourly wind speed for Riyadh site

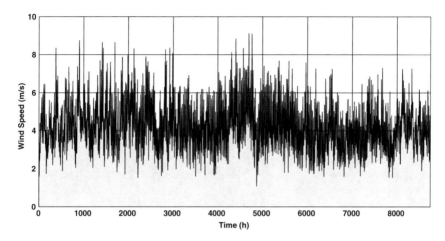

Fig. A.4 Hourly wind speed for Qaisumah site

Appendix B
Solar Radiation for Sites Under Study

See Figs. B.1, B.2, B.3 and B.4.

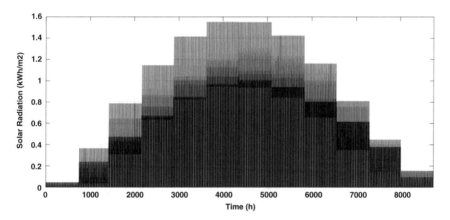

Fig. B.1 Hourly tilted solar radiation for Dhahran site

M. Abdelaziz Mohamed and A.M. Eltamaly, *Modeling and Simulation of Smart Grid Integrated with Hybrid Renewable Energy Systems*, Studies in Systems, Decision and Control 121, DOI 10.1007/978-3-319-64795-1

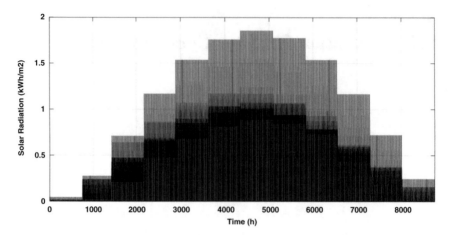

Fig. B.2 Hourly tilted solar radiation for Dhalm site

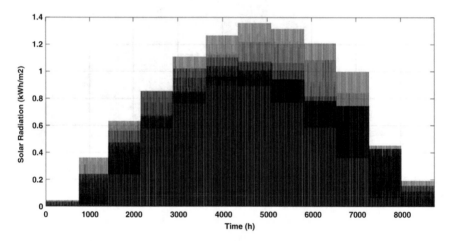

Fig. B.3 Hourly tilted solar radiation for Riyadh site

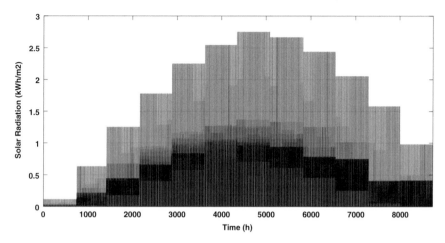

Fig. B.4 Hourly tilted solar radiation for Qaisumah site

References

1. Eltamaly, A. M., Mohamed, M. A., Al-Saud, M. S., & Alolah, A. I. (2016). Load management as a smart grid concept for sizing and designing of hybrid renewable energy systems. Engineering Optimization, 1–16.
2. Zhang, P., Li, F., & Bhatt, N. (2010). Next-generation monitoring, analysis, and control for the future smart control center. IEEE Transactions on Smart Grid, 1(2), 186–192.
3. Güngör, V. C., Sahin, D., Kocak, T., Ergüt, S., Buccella, C., Cecati, C., et al. (2011). Smart grid technologies: Communication technologies and standards. IEEE Transactions on Industrial Informatics, 7(4), 529–539.
4. Rahimi, F., & Ipakchi, A. (2010). Overview of demand response under the smart grid and market paradigms. In Proceedings of the IEEE PES Innovative Smart Grid Technologies (ISGT) (pp. 1–7).
5. Motamedi, A., Zareipour, H., & Rosehart, W. D. (2012). Electricity price and demand forecasting in smart grids. IEEE Transactions on Smart Grid, 3(2), 664–674.
6. Mohamed, M. A., Eltamaly, A. M., & Alolah, A. I. (2016). PSO-based smart grid application for sizing and optimization of hybrid renewable energy systems. PloS one, 11(8), e0159702.
7. Phuangpornpitak, N., & Tia, S. (2011). Feasibility study of wind farms under the Thai very small scale renewable energy power producer (VSPP) program. Energy Procedia, 9, 159–170.
8. Phuangpornpitak, N., & Tia, S. (2013). Opportunities and challenges of integrating renewable energy in smart grid system. Energy Procedia, 34, 282–290.
9. Gaviano, A., Weber, K., & Dirmeier, C. (2012). Challenges and integration of PV and wind energy facilities from a smart grid point of view. Energy Procedia, 25, 118–125.
10. Ayompe, L. M., Duffy, A., McCormack, S. J., & Conlon, M. (2010). Validated real-time energy models for small-scale grid-connected PV-systems. Energy, 35(10), 4086–4091.
11. Crow, M. L., McMillin, B., Wang, W., & Bhattacharyya, S. (2010). Intelligent energy management of the FREEDM system. In Proceedings of the IEEE PES General Meeting (pp. 1–4).
12. Kohsri, S., & Plangklang, B. (2011). Energy management and control system for smart renewable energy remote power generation. Energy Procedia, 9, 198–206.
13. Farhangi, H. (2010). The path of the smart grid. IEEE Power and Energy Magazine, 8(1), 18–28.
14. Borowy, B. S., & Salameh, Z. M. (1996). Methodology for optimally sizing the combination of a battery bank and PV array in a wind/PV hybrid system. IEEE Transactions on Energy Conversion, 11(2), 367–375.
15. Chedid, R., & Rahman, S. (1997). Unit sizing and control of hybrid wind-solar power systems. IEEE Transactions on Energy Conversion, 12(1), 79–85.
16. Belfkira, R., Zhang, L., & Barakat, G. (2011). Optimal sizing study of hybrid wind/PV/diesel power generation unit. Solar Energy, 85(1), 100–110.

© Springer International Publishing AG 2018
M. Abdelaziz Mohamed and A.M. Eltamaly, *Modeling and Simulation of Smart Grid Integrated with Hybrid Renewable Energy Systems*, Studies in Systems, Decision and Control 121, DOI 10.1007/978-3-319-64795-1

17. Celik, A. N. (2003). Techno-economic analysis of autonomous PV-wind hybrid energy systems using different sizing methods. *Energy Conversion and Management, 44*(12), 1951–1968.

18. Diaf, S., Notton, G., Belhamel, M., Haddadi, M., & Louche, A. (2008). Design and techno-economical optimization for hybrid PV/wind system under various meteorological conditions. *Applied Energy, 85*(10), 968–987.

19. Kaabeche, A., Belhamel, M., & Ibtiouen, R. (2011). Sizing optimization of grid-independent hybrid photovoltaic/wind power generation system. *Energy, 36*(2), 1214–1222.

20. Yang, H., Lu, L., & Zhou, W. (2007). A novel optimization sizing model for hybrid solar-wind power generation system. *Solar energy, 81*(1), 76–84.

21. Kellogg, W. D., Nehrir, M. H., Venkataramanan, G., & Gerez, V. (1998). Generation unit sizing and cost analysis for stand-alone wind, photovoltaic, and hybrid wind/PV systems. *IEEE Transactions on Energy Conversion, 13*(1), 70–75.

22. Hocaoğlu, F. O., Gerek, Ö. N., & Kurban, M. (2009). A novel hybrid (wind-photovoltaic) system sizing procedure. *Solar Energy, 83*(11), 2019–2028.

23. Nandi, S. K., & Ghosh, H. R. (2009). A wind-PV-battery hybrid power system at Sitakunda in Bangladesh. *Energy Policy, 37*(9), 3659–3664.

24. Barley, C. D., Lew, D. J., & Flowers, L. T. (1997). *Sizing wind/photovoltaic hybrids for households in Inner Mongolia.* Golden, CO: National Renewable Energy Laboratory.

25. Eltamaly, A. M., Addoweesh, K. E., Bawa, U., & Mohamed, M. A. (2014). Economic modeling of hybrid renewable energy system: A case study in Saudi Arabia. *Arabian Journal for Science and Engineering, 39*(5), 1–13.

26. Bao, Y., Chen, X., Wang, H., & Wang, B. (2013). Genetic algorithm based optimal capacity allocation for an independent wind/pv/diesel/battery power generation system. *Journal of Information and Computational Science, 10*(14), 4581–4592.

27. Katsigiannis, Y. A., Georgilakis, P. S., & Karapidakis, E. S. (2010). Genetic algorithm solution to optimal sizing problem of small autonomous hybrid power systems. *Lecture Notes in Artificial Intelligence, 6040*, 327–332.

28. Yang, H., Wei, Z., & Chengzhi, L. (2009). Optimal design and techno-economic analysis of a hybrid solar-wind power generation system. *Applied Energy, 86*(2), 163–169.

29. Mellit, A. (2009). Application of genetic algorithm and neural network for sizing of hybrid photovoltaic wind power generation (HPVWPG) systems. *International Journal of Renewable Energy Technology, 1*(2), 139–154.

30. Boonbumroong, U., Pratinthong, N., Thepa, S., Jivacate, C., & Pridasawas, W. (2011). Particle swarm optimization for AC-coupling stand-alone hybrid power systems. *Solar Energy, 85*(3), 560–569.

31. Bashir, M., & Sadeh, J. (2012, May) Optimal sizing of hybrid wind/photovoltaic/battery considering the uncertainty of wind and photovoltaic power using Monte Carlo. In *Proceedings of the 11th IEEE International Conference on Environment and Electrical Engineering (EEEIC)* (pp. 1081–1086).

32. Kaviani, A. K., Riahy, G. H., & Kouhsari, S. M. (2009). Optimal design of a reliable hydrogen-based stand-alone wind/PV generating system, considering component outages. *Renewable Energy, 34*(11), 2380–2390.

33. Ardakani, F. J., Riahy, G., & Abedi, M. (2010, May) Design of an optimum hybrid renewable energy system considering reliability indices. In *Proceedings of the18th IEEE Iranian Conference on Electrical Engineering (ICEE)* (pp. 842–847).

34. Bashir, M., & Sadeh, J. (2012, May) Size optimization of new hybrid stand-alone renewable energy system considering a reliability index. In *Proceedings of the 11th IEEE International Conference on Environment and Electrical Engineering (EEEIC)* (pp. 989–994).

35. Askarzadeh, A., & dos Santos Coelho, L. (2015). A novel framework for optimization of a grid independent hybrid renewable energy system: A case study of Iran. *Solar Energy, 112,* 383–396.

36. Mohammadi, M., Hosseinian, S. H., & Gharehpetian, G. B. (2012). Optimization of hybrid solar energy sources/wind turbine systems integrated to utility grids as microgrid (MG) under pool/bilateral/hybrid electricity market using PSO. *Solar Energy, 86*(1), 112–125.

37. Amer, M., Namaane, A., & M'Sirdi, N. K. (2013). Optimization of hybrid renewable energy systems (HRES) using PSO for cost reduction. *Energy Procedia, 42,* 318–327.

38. Hakimi, S. M., & Moghaddas-Tafreshi, S. M. (2009). Optimal sizing of a stand-alone hybrid power system via particle swarm optimization for Kahnouj area in South-East of Iran. *Renewable energy, 34*(7), 1855–1862.

39. Lee, T. Y., & Chen, C. L. (2009). Wind-photovoltaic capacity coordination for a time-of-use rate industrial user. *IET Renewable Power Generation, 3*(2), 152–167.

40. Wang, L., & Singh, C. (2009). Multicriteria design of hybrid power generation systems based on a modified particle swarm optimization algorithm. *IEEE Transactions on Energy Conversion, 24*(1), 163–172.

41. Cui, T., Goudarzi, H., Hatami, S., Nazarian, S., & Pedram, M. (2012). Concurrent optimization of consumer's electrical energy bill and producer's power generation cost under a dynamic pricing model. In *Proceedings of the IEEE PES Innovative Smart Grid Technologies (ISGT)* (pp. 1–6).

42. Conejo, A. J., Morales, J. M., & Baringo, L. (2010). Real-time demand response model. *IEEE Transactions on Smart Grid, 1*(3), 236–242.

43. Caron, S., & Kesidis, G. (2010). Incentive-based energy consumption scheduling algorithms for the smart grid. In *Proceedings of the IEEE International Conference on Smart Grid Communications (SmartGridComm)* (pp. 391–396).

44. Kishore, S., & Snyder, L. V. (2010). Control mechanisms for residential electricity demand in smartgrids. In *Proceedings of the IEEE International Conference on Smart Grid Communications (SmartGridComm)* (pp. 443–448).

45. Mohsenian-Rad, A. H., & Leon-Garcia, A. (2010). Optimal residential load control with price prediction in real-time electricity pricing environments. *IEEE Transactions on Smart Grid, 1*(2), 120–133.

46. Ghosh, S., Kalagnanam, J., Katz, D., Squillante, M., Zhang, X., & Feinberg, E. (2010). Incentive design for lowest cost aggregate energy demand reduction. In *Proceedings of the IEEE International Conference on Smart Grid Communications (SmartGridComm)* (pp. 519–524).

47. Lee, J., & Park, G. L. (2013). Power load distribution for wireless sensor and actuator networks in smart grid buildings. *International Journal of Distributed Sensor Networks, 2013.*

48. Neill, D. O., & Levorato, M. Goldsmith, A., & Mitra, U. (2010). Residential demand response using reinforcement learning. In *Proceedings of the IEEE International Conference on Smart Grid Communications (SmartGridComm)* (pp. 409–414).

49. Bakker, V., Bosman, M. G. C., Molderink, A., Hurink, J. L., & Smit, G. J. M. (2010). Demand side load management using a three step optimization methodology. In *Proceedings of the IEEE International Conference on Smart Grid Communications (SmartGridComm)* (pp. 431–436).

50. Saber, A. Y., & Venayagamoorthy, G. K. (2011). Plug-in vehicles and renewable energy sources for cost and emission reductions. *IEEE Transactions on Industrial Electronics, 58*(4), 1229–1238.

51. Setiawan, A. A., Zhao, Y., & Nayar, C. V. (2009). Design, economic analysis and environmental considerations of mini-grid hybrid power system with reverse osmosis desalination plant for remote areas. *Renewable Energy, 34*(2), 374–383.

52. Wang, C., & Nehrir, M. H. (2008). Power management of a stand-alone wind/photovoltaic/fuel cell energy system. *IEEE Transactions on Energy Conversion, 23*(3), 957–967.

53. Eltamaly, A. M., & Mohamed, M. A. (2014). A novel design and optimization software for autonomous PV/wind/battery hybrid power systems. *Mathematical Problems in Engineering*, 637174, 1–16.

54. Eltamaly, A. M., Addoweesh, K. E., Bawah, U., & Mohamed, M. A. (2013). New software for hybrid renewable energy assessment for ten locations in Saudi Arabia. *Journal of Renewable and Sustainable Energy, 5*(3), 033126.

55. Bernal-Agustín, J. L., & Dufo-López, R. (2009). Simulation and optimization of stand-alone hybrid renewable energy systems. *Renewable and Sustainable Energy Reviews, 13*(8), 2111–2118.

56. Sreeraj, E. S., Chatterjee, K., & Bandyopadhyay, S. (2010). Design of isolated renewable hybrid power systems. *Solar Energy, 84*(7), 1124–1136.

57. Lun, I. Y., & Lam, J. C. (2000). A study of Weibull parameters using long-term wind observations. *Renewable Energy, 20*(2), 145–153.

58. Mohamed, M. A., Eltamaly, A. M., & Alolah, A. I. (2015). Sizing and techno-economic analysis of stand-alone hybrid photovoltaic/wind/diesel/battery power generation systems. *Journal of Renewable and Sustainable Energy, 7*(6), 063128.

59. Eltamaly, A. M., & Mohamed, M. A. (2016). A novel software for design and optimization of hybrid power systems. *Journal of the Brazilian Society of Mechanical Sciences and Engineering, 38*(4), 1299–1315.

60. Habib, M. A., Said, S. A. M., El-Hadidy, M. A., & Al-Zaharna, I. (1999). Optimization procedure of a hybrid photovoltaic wind energy system. *Energy, 24*(11), 919–929.

61. Skoplaki, E., Boudouvis, A. G., & Palyvos, J. A. (2008). A simple correlation for the operating temperature of photovoltaic modules of arbitrary mounting. *Solar Energy Materials and Solar Cells, 92*(11), 1393–1402.

62. Mohamed, M. A., Eltamaly, A. M., & Alolah, A. I. (2017). Swarm intelligence-based optimization of grid-dependent hybrid renewable energy systems. *Renewable and Sustainable Energy Reviews, 77*, 515–524.

63. Yang, H., Zhou, W., Lu, L., & Fang, Z. (2008). Optimal sizing method for stand-alone hybrid solar-wind system with LPSP technology by using genetic algorithm. *Solar Energy, 82*(4), 354–367.

64. Ismail, M. S., Moghavvemi, M., & Mahlia, T. M. I. (2013). Techno-economic analysis of an optimized photovoltaic and diesel generator hybrid power system for remote houses in a tropical climate. *Energy Conversion and Management, 69*, 163–173.

65. Dufo-López, R., Bernal-Agustín, J. L., Yusta-Loyo, J. M., Domínguez-Navarro, J. A., Ramírez-Rosado, I. J., Lujano, J., et al. (2011). Multi-objective optimization minimizing cost and life cycle emissions of stand-alone PV-wind-diesel systems with batteries storage. *Applied Energy, 88*(11), 4033–4041.

66. Nema, P., Nema, R. K., & Rangnekar, S. (2009). A current and future state of art development of hybrid energy system using wind and PV-solar: A review. *Renewable and Sustainable Energy Reviews, 13*(8), 2096–2103.

67. Diaf, S., Belhamel, M., Haddadi, M., & Louche, A. (2008). Technical and economic assessment of hybrid photovoltaic/wind system with battery storage in Corsica island. *Energy Policy, 36*(2), 743–754.

68. Focuseconomics, Interest Rate in Saudi Arabia. Available at: http://www.focus-economics.com/country-indicator/saudi-arabia/interest-rate. Accessed 27 January 2016.

69. Kaabeche, A., Belhamel, M., & Ibtiouen, R. (2010). Optimal sizing method for stand-alone hybrid PV/wind power generation system. *Revue des Energies Renouvelables (SMEE'10) Bou Ismail Tipaza*, 205–213.

70. Diaf, S., Diaf, D., Belhamel, M., Haddadi, M., & Louche, A. (2007). A methodology for optimal sizing of autonomous hybrid PV/wind system. *Energy Policy, 35*(11), 5708–5718.

71. Lazou, A. A., & Papatsoris, A. D. (2000). The economics of photovoltaic stand-alone residential households: a case study for various European and Mediterranean locations. *Solar Energy Materials and Solar Cells, 62*(4), 411–427.

72. Nelson, D. B., Nehrir, M. H., & Wang, C. (2005). Unit sizing of stand-alone hybrid wind/PV/fuel cell power generation systems. In *Proceedings of IEEE Power Engineering Society General Meeting* (pp. 2116–2122).

73. Navaeefard, A., Tafreshi, S., & Maram, M. (2010). Distributed energy resources capacity determination of a hybrid power system in electricity market. In *Proceedings of the 25th International Power System Conference, (PSC)* (pp. 1–9).

74. Erdinc, O., & Uzunoglu, M. (2012). Optimum design of hybrid renewable energy systems: overview of different approaches. *Renewable and Sustainable Energy Reviews, 16*(3), 1412–1425.

75. Saudi Arabia Inflation Rate, Trading Economics. Available at: http://www.tradingeconomics.com/saudi-arabia/inflation-cpi. Accessed 27 January 2016.

76. Shata, A. A., & Hanitsch, R. (2006). The potential of electricity generation on the east coast of Red Sea in Egypt. *Renewable Energy, 31*(10), 1597–1615.

77. Ucar, A., & Balo, F. (2009). Evaluation of wind energy potential and electricity generation at six locations in Turkey. *Applied Energy, 86*(10), 1864–1872.

78. Kaabeche, A., & Ibtiouen, R. (2014). Techno-economic optimization of hybrid photovoltaic/wind/diesel/battery generation in a stand-alone power system. *Solar Energy, 103*, 171–182.

79. ISH TECHNOLOGY, Power and Energy Technology. Available at: https://technology.ihs.com/Research-by-Market/450473/power-energy. Accessed 29 November 2015.

80. Worldatlas, Saudi Arabia Available at: http://www.worldatlas.com/webimage/countrys/asia/sa.htm. Accessed 30 January 2016.

81. Renewable Resource Atlas, Maps and Graphs. Available at: http://rratlas.kacare.gov.sa/RRMMDataPortal/en/MapTool. Accessed 30 January 2016.

82. Wind energy database, The Wind Power. Available at: http://www.thewindpower.net/index.php. Accessed 30 January 2016.

83. Eroglu, M., Dursun, E., Sevencan, S., Song, J., Yazici, S., & Kilic, O. (2011). A mobile renewable house using PV/wind/fuel cell hybrid power system. *International journal of hydrogen energy, 36*(13), 7985–7992.

84. Etamaly, A. M., Mohamed, M. A., & Alolah, A. I. (2015, August). A smart technique for optimization and simulation of hybrid photovoltaic/wind/diesel/battery energy systems. In *IEEE International Conference on Smart Energy Grid Engineering (SEGE)*, 1–6.

85. Mohamed, M. A., Eltamaly, A. M., Farh, H. M., & Alolah, A. I. (2015, August). Energy management and renewable energy integration in smart grid system. In *Proceedings of IEEE International Conference on Smart Energy Grid Engineering (SEGE)* (pp. 1–6).

86. Eltamaly, A. M., Mohamed, M. A., & Alolah, A. I. (2016). A novel smart grid theory for optimal sizing of hybrid renewable energy systems. *Solar Energy*, 124, 26–38.

87. Fadaee, M., & Radzi, M. A. M. (2012). Multi-objective optimization of a stand-alone hybrid renewable energy system by using evolutionary algorithms: A review. *Renewable and Sustainable Energy Reviews, 16*(5), 3364–3369.

88. Bai, Q. (2010). Analysis of particle swarm optimization algorithm. *Computer and information science, 3*(1), 180–184.

89. Kennedy, J., & Ebethart, R. (1995). Particle swarm optimization. In *Proceedings of IEEE International Conference on Neural Networks* (pp. 1942–1948).

90. Ardakani, A. J., Ardakani, F. F., & Hosseinian, S. H. (2008). A novel approach for optimal chiller loading using particle swarm optimization. *Energy and Buildings, 40*(12), 2177–2187.

91. Wang, L., & Singh, C. (2007, April). PSO-based multi-criteria optimum design of a grid-connected hybrid power system with multiple renewable sources of energy. In *Proceedings of IEEE Swarm Intelligence Symposium* (pp. 250–257).

92. Das, S., Abraham, A., & Konar, A. (2008). Particle swarm optimization and differential evolution algorithms: Technical analysis, applications and hybridization perspectives. In *Advances of Computational Intelligence in Industrial Systems* (pp. 1–38).

93. Sinha, S., & Chandel, S. S. (2014). Review of software tools for hybrid renewable energy systems. *Renewable and Sustainable Energy Reviews, 32*, 192–205.

94. Mathworks. Available at: http://www.mathworks.com/help/distcomp/parallel-pools.html. Accessed 29 December 2015.

Printed in the United States
By Bookmasters